Towards Revealing the Origin of Life

Kenji Ikehara

Towards Revealing the Origin of Life

Presenting the GADV Hypothesis

Springer

Kenji Ikehara
G&L Kyousei Institute
Nara, Japan

ISBN 978-3-030-71086-6   ISBN 978-3-030-71087-3 (eBook)
https://doi.org/10.1007/978-3-030-71087-3

This Springer imprint is published by the registered company Springer Nature Switzerland AG
The registered company address is: Gewerbestrasse 11, 6330 Cham, Switzerland

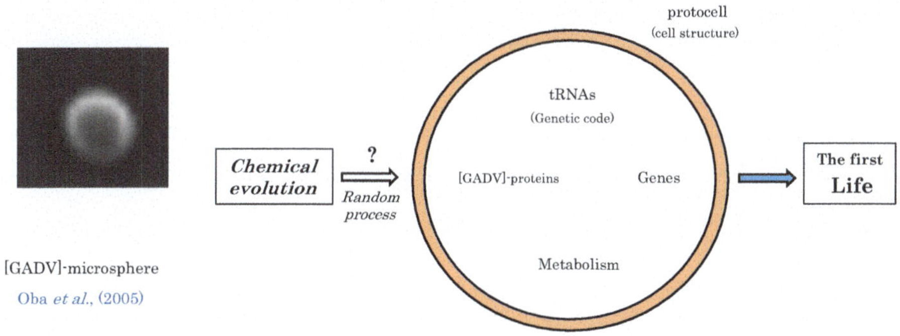

(**Left**) Scanning electron microscope image of a [GADV]-microsphere, which I have supposed to as a protocell. (**Right**) Modern life is living under the fundamental life system composed of six members: protein, cell structure, metabolism, tRNA, genetic code, and gene. [GADV] or GADV means four amino acids, Gly [G], Ala [A], Asp [D], and Val [D]. A cartoon (right) shows the six members, which are indispensable to the emergence of life. However, all the first six members must be created through the respective random processes. How were the six members created through the random processes, which only occurred on the primitive Earth? The objective of this book is to show the steps from chemical evolution to the emergence of life through the creation of the six members.

How could life emerge through random process on the primitive Earth? I have aimed at giving an answer for solving the riddle of the origin of life from my own viewpoints of protein 0th-order structure and [GADV]-protein world hypothesis (GADV hypothesis) in this book.

Foreword

I have been following Kenji Ikehara' work for many years, my attraction being based on two main features: firstly, that he was starting all his considerations based on the firm ground of the four basic amino acids G, A, D, and V and their polycondensates, which were abundant in prebiotic time; secondly, that he, simultaneously to the focus on polypeptides, was considering also their genetic correlates. This double view, or, better, the view that the two affairs must be stringed together, located Ikehara for me in a different position with respect to most of the researchers on the origin of life, who generally were working within the tiny garden of a very localized perspective – losing then the view of the vast forest of life at large.

Ikehara starts with the basic assumption that in order to understand the origin of life, you have to start from the "six members," namely proteins, cell structure, metabolism, tRNA, genetic code, and gene. And in fact, he devotes entire chapters of the book to each of these members.

The book consists of ten chapters, which are structurally well organized – each one with a summary at the beginning, then the exposition of the chapter's main hypothesis, then a section on the weakness of such a hypothesis, a section on its strength, and finally a discussion, and each chapter concludes with a rich reference list.

Of course, in each chapter, the main hero is GADV. Thus, the primordial cell is a GADV microsphere (Chap. 4), assuming that the modern cell is the evolutionary product of a primitive cell derived from such a microsphere and that the metabolism has its origin in the primordial catalytic activity of GADV (Chap. 5), whereby the modern metabolism originated from the three simple organic compounds glyoxylate, glyceraldehyde, and pyruvate, all within the more general view that proteins, not genes, would play the central metabolic game, because proteins exhibit almost all biological functions and because genes and genetic code have been only supporting protein synthesis through the life system.

In general, we can say that this is a rather theoretical book, although Ikehara refers at times to his studies on the mature proteins of bacteria.

In describing the GADV world, Ikehara also builds his own dictionary – a terminology which, in my opinion, may be important for the field at large. Thus, he distinguishes between immature and mature proteins, also introducing the concept of "protein 0th-order structure" (in which immature GADV-protein could be produced by random joining of GADV amino acids); he also considers two sides of the origin of genes: firstly, the origin of gene-I (how was the gene created before the origin of life?) and, secondly, of gene-II (how did we go from there to the present?).

There are in fact three chapters (Chaps. 6, 7, 8) devoted to the origin of genes, and most of the readers may enjoy Chap. 6 on the origin of tRNA. All these three chapters are very detailed – and not easy, in the sense that you have really to concentrate hard in order to follow the reasoning of the author. And here and there, if you are a malicious reader, you may detect small pieces of "magics" – hypothetical events that must have happened, otherwise the entire story does not flow – like the origination of the first nonspecific anticodon stem-loop tRNA or the point at which the GADV microsphere acquired double-stranded $(GNC)_n$ genes.

Quite in general, some piece of magics is always present in the various hypotheses of the origin of life – Do you remember Crick's "frozen accident"? A way to disguise magics in almost scientific wording – not to mention the miracles of the RNA world.

While the three chapters on genes are very detailed, the passage from the macromolecular component to life is, in comparison, rather quick. There is an entire chapter (Chap. 9) with the specific title "The Origin of Life", and here one may miss the literature on the synthetic biology of the minimal cells, as developed first at the ETH Zurich in 1985 and later by many others (the project of inserting proteins and/or nucleic acids in vesicles, with the idea of constructing by a bottom-up approach "minimal cell" , having the minimal amount of proteins and nucleic acids to behave like a real cell – an approach which I now, after having worked for years on it, consider impracticable).

In his ample theoretical framework, Ikehara also asks many provocative questions, actually each chapter contains five to six such questions, and people in the field, regardless of their acceptance or not in the GADV hypothesis, should take these questions seriously and try to provide an answer, or at least make a mental note of them.

The author also devotes much energy to counteract the alternative hypotheses on the origin of life (Chap. 9, and see also the Introduction), and I agree with him, for example, that the gene-centered view, which is at the basis of the RNA world, is actually hindering the progress in the field in a kind of immobile crystallization.

We mentioned earlier, without going in the details, some little pieces of magic inherent in each hypothesis on the origin of life. The RNA world is the most striking example of that: it starts right away with the magical appearance of a self-replicating RNA, and this, to make sense from the chemical point of view, must be a weighed amount –to ensure that its concentration in the small warm pond would be high enough. This is in order to permit the interaction of RNA dimers (otherwise who is replicating what?), whereby the catalytic prebiotic RNA should also catalyze the insertion in the growing new chain of all four different nucleotides (second miracle) – which, by the way, must be anyway present in great excess. And then this replicated RNA should transform itself (thanks to how many mutations?) into a ribozyme which catalyzes all possible different peptide bonds (third miracle) to make proteins. Actually, the fact that all this RNA world is taken seriously by a generation of scientists appears to me a social mass psychology problem. In this sense, let me explain the more realistic GADV world.

The other aspect to be stressed in the GADV approach is that the origin of life is seen as a bottom-up unitarian process, see Figs. 9.11, 9.12, 4.2, 9.2, Fig. 2.4, and several others: an uninterrupted series of events, starting with the GADV polypeptides and arriving, step by step, to life (although surprising, Ikehara at a certain point also talks about top-down approach, see Fig. 10.3 – maybe just a question of semantics). Personally, but this is again a personal prejudice, I don't like the view that life developed on Earth as a single process, namely from the first prebiotic molecules and going up in complexity and functionality till the end – the cell, or better, a colony of cells. I rather prefer the notion of contingency à la Gould, whereby several causally independent events casually interacted with each other giving rise to a complexity which could not be obtained by a logical continuous flow – one of the main reasons why actually we do not comprehend the origin of life with our logical mind and why we have to introduce some magics in order to explain it.

Just because of the necessity of this magics, in general, I personally still consider the origin of life a mystery – meaning by that a problem which has not yet found and seem unable to find a satisfactory scientific explanation, neither conceptually, even less experimentally. If the emergence of life was really a spontaneous "easy event" – probably occurred in many places and many times simultaneously on Earth – then why the best chemical minds have not been able to, in 70 years of investigation, make a single prokaryotic cell colony in the lab?

On the other hand, I don't want that this skeptic statement of mine be taken, especially by young people, as an invitation to abandon or diminish the research on the field. There is nothing more exciting than to face a deep mystery – actually, if I were only 50 years younger, I would jump again in the lab to continue the work.

But let me conclude going back to the book by Ikehara. A book can be magistral even if you do not accept all of it, it is magistral when it opens your mind to a wider horizon, enriched by new and broader questions. Ikehara's book does that. In addition, it must be honored as the account of an entire life devoted to one of the most profound questions of our planet.

Professor Emeritus ETH Zurich, Pier Luigi Luisi
Zürich, Switzerland

Preface

When we see various kinds of organisms, which prosperously inhabit the present Earth, we think how such beautiful organisms emerged on this planet. To answer this question, many researchers have tried to solve the riddle of life's origin. Oparin proposed the coacervate hypothesis based on formation of cell-like structure, and Fox proposed the proteinoid microsphere hypothesis based on the formation of quite small spherical cell-like structure. Further, Miller confirmed that organic compounds such as amino acids could be produced from simple inorganic compounds as CH_4, H_2O, and NH_3 by electric discharge experiments. They naturally considered that life must emerge with proteins, because their experiments were carried out before or around the discovery of double-stranded DNA by Watson and Crick.

On the other hand, mechanisms of various important biological matters, such as replication of DNA or genetic material, expression of gene written as base sequence onto DNA, and relation between mutations in gene accompanied by base substitutions and biological evolution, were clarified one after another upon discovering the structure of double-stranded DNA. It can be easily understood that many biological phenomena are intimately related to DNA or gene. Along with that, gene-centered idea became strong as many researchers recognized the importance of self-replication of DNA. Certainly, it can be easily supposed that no biological activity can be expressed without gene and the extant organisms on Earth would not have evolved without it.

The origin of life was reconsidered in such an atmosphere. From the results, it was understood that there exists the so called "chicken-egg relationship" between gene and protein. This led to a situation where many persons considered that life did not emerge without gene and also could not emerge without protein. Then, ribozyme, which has genetic information similar to DNA and expresses catalytic activity similar to enzyme, was discovered. The discovery of ribozyme gave birth to a new theory on the origin of life, that is, the RNA world hypothesis. I also considered that the RNA world hypothesis is quite interesting and the riddle of life's origin might be solved by this hypothesis, when I read the paper, which was published in Nature by W. Gilbert. I boastfully introduced the RNA world hypothesis to many students in my lecture. However, it must be stated that the origin of life unfortunately remains unsolved as of now irrespective of strenuous efforts by many researchers for about 35 years since the proposition of the RNA world hypothesis.

I began researching the origin of modern genes, how entirely new genes have been created on the present Earth, but not on the origin of life, in a laboratory in the Department of Chemistry, Faculty of Science, Nara Women's University, with my students about 30 years ago. Consequently, I reached one conclusion, "GC-NSF(a) hypothesis," assuming that an entirely new gene has been created from a nonstop frame on antisense strand of a GC-rich gene (GC-NSF(a)). The GC-NSF(a) hypothesis triggered the GNC-SNS primitive genetic code hypothesis and [GADV]-protein world hypothesis or GADV hypothesis on the origin of life. Since then, I have continued to confirm validity of the GADV hypothesis, and I am now convinced that the GADV hypothesis is correct as it explains the steps from chemical evolution to the emergence of life. Please consider well my ideas about the GADV hypothesis described in this book and send me your thoughts, comments, questions, consenting opinions, or anything else through email, and I will respond to them as much as possible. I hope that discussion about the origin of life will be thrust into the spotlight triggered by the publication of this book.

Nara, Japan
November 8, 2020

Kenji Ikehara

Acknowledgments

I am very grateful to Dr. Tadashi Oishi (G&L Kyosei Institute, emeritus professor at Nara Women's University) for encouragement throughout my research on the origin and evolution of the fundamental life system, especially after my retirement from Nara Women's University. I also thank emeritus professors Dr. Yoshiomi Takagi and Dr. Terue Harumoto at Nara Women's University for their warmest supports to my research. I am thankful to my students, who worked hard to discover the origin of life in my laboratory of biochemistry, Department Chemistry, Faculty of Science, Nara Women's University, because my investigation on the origin of life never progressed well without their endeavors.

Finally, I am very grateful to Professor Pier Luigi Luisi for the cordial and appropriate Foreword. I would like to express my gratitude to Dr. Massimo di Giulio for precious comments on my rough manuscript and especially to the editor of this book, Dr. Gonzalo Cordova, for his valuable advises and assistance in publishing it.

Contents

About the Author

Kenji Ikehara is an emeritus professor at Nara Women's University and also a fellow of International Institute for Advanced Studies (IIAS) of Japan. He graduated from the Department of Industrial Chemistry, Faculty of Engineering, Kyoto University, in 1968 and received his B.Eng. (1968), and successively M. Eng. (1970) and D. Eng. (1976) degrees from Kyoto University. He worked as a research associate in Faculty of Science, the University of Tokyo, and successively as an associate professor in the Faculty of Science, Nara Women's University. He further got promoted to professor at the same university. He later became dean of the Faculty of Science, Nara Women's University, and director of Nara Study Center of the Open University of Japan. He is engaged in studies on sporulation initiation of *Bacillus subtilis* and on the origins and evolutionary processes of microbial genes, the genetic code, and proteins and life for around 30 years at Nara Women's University. Consequently, he has proposed the GC-NSF(a) hypothesis on the origin of genes, GNC-SNS hypothesis on the genetic code, protein 0th-order structure hypothesis on the origin of protein, and [GADV]-protein world hypothesis (GADV hypothesis) on the origin of life. He also served as local chair of the International Conference, Origin 2014, which was held at Nara in 2014.

Abbreviations

GADV Hypothesis Related Abbreviations

[GADV], GADV	Four amino acids: glycine, Gly [G]; alanine, Ala [A]; aspartic acid, Asp [D]; and valine, Val [V]
GNC	N means four bases: adenine, A; uracil or thymine, U/T; guanine, G; or cytosine, C. GNC = GAC, GUC, GGC, and GCC
SNS	S means guanine, G or cytosine; C. SNS = (16 codons/anticodons) GAC, GUC, GGC, GCC, GAG, GUG, GGG, GCG, CAC, CUC, CGC, CCC, CAG, CUG, CGG, and CCG
GC-NSF(a)	Nonstop frame on antisense strand of a GC-rich gene
AntiC-SL	Anticodon stem-loop
EntNew	Entirely new
GoGaPyr	glyoxylate (Go), glyceraldehyde (Ga) and pyruvate (Pyr)
ss, ds	single-stranded, double-stranded
iPP protein	immature pluripotent protein
OS	ordered sequence

Others

TCA cycle	Tricarboxylic acid
Acetyl-CoA	Acetyl-coenzyme A
ATP	Adenosine triphosphate
PRPP	Phosphoribosyl-pyrophosphate
ARS	aminoacyl-tRNA synthetase
Diapr	L-2,3-diaminopropionic acid
SELEX	Systematic evolution of ligands by exponential enrichment
CLM	Coenzyme-like molecule

Chapter 1
Introduction

Abstract Diverse organisms are living on the present Earth. This clearly indicates that life emerged on the primitive Earth in the remote past. Therefore, many researchers have tried to solve the riddle of the origin of life. However, unfortunately, the riddle has been unsolved despite the strenuous efforts of many researchers. Then, first, the reasons, why the origin of life remains unsolved still now, are discussed from viewpoints of research methodologies and so on. On the other hand, it is necessary to make clear how the three main members in the fundamental life system, or gene, tRNA (genetic code) and protein, were established on the primitive Earth in order to solve the riddle. In addition, it must be reasonably explained that the three members were created through the respective random processes, because the three members never be designed in advance. On the contrary, it was found that the steps from chemical evolution to the emergence of life can be rationally explained according to [GADV]-protein world hypothesis or GADV hypothesis on the origin of life, which I have proposed. Therefore, I have now considered that the establishment process of the fundamental life system could be comprehensively explained under one concept obtained from the original viewpoint, one of "protein 0th-order structures" or "[GADV]-amino acids". Then, the GADV hypothesis is introduced in short for readers to survey my research results before taking up the main subjects described in the respective Chapters of this book. Furthermore, the reasons, why I could reach the novel idea, and an overview of this book are also described in this chapter.

Keywords Research methodology · Bottom-up approach · The fundamental life system

Diverse organisms are living on the present Earth. The origin of life is one of the interesting and important research fields, how the life emerged on the primitive Earth about 4 billion years ago. The first life evolved to extant organisms. Consequently, diverse organisms, like as mammals including human beings, fishes, insects, flowers and even bacteria, are inhabiting in various environments on the present Earth. We, human beings, have been interested in the origin of life, how life

K. Ikehara, *Towards Revealing the Origin of Life*,
https://doi.org/10.1007/978-3-030-71087-3_1

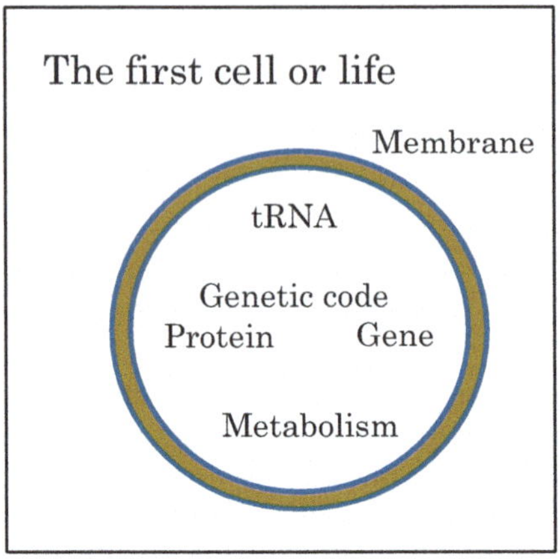

Fig. 1.1 What is life?
1. Individuality for proliferation. 2. Taking in nutrients (organic compounds). 3. Metabolizing the nutrients. 4. Self-reproduction. 5. Evolution. 6. Living under the fundamental life system composed of protein, cell structure (membrane), metabolism, tRNA, genetic code and gene. Elucidating the origin of life is to understand how the six members composing the fundamental life system were created through random processes in what order and consequently how the first life emerged

arose on this planet, for many years. For example, Arithtoteles, Oparin, Fox and Miller tackled the riddle of the origin of life. After them, many researchers also have tried to solve the riddle, because the clues for knowing how and from where we have come to the present Earth, might be found through the investigation on the origin of life. Thus, the study on the origin of life has been also a long journey until now to understand "what is life?", "why are we here?", "for what purpose are we living?", "What is the meaning of my life?" and so on (Fig. 1.1).

However, unfortunately, the riddle of the origin of life has been unsolved despite the strenuous efforts of many researchers still now. It is obvious even from information retrieved recently from the Internet that the research is still in the middle of a long journey for solving the origin of life. It seems to me that many researchers working in the field of the life-origin believe that the origin of life could be proved based on the RNA world hypothesis (Gilbert 1986) considering that life emerged from RNA world, if nucleotides and RNA could be synthesized with prebiotic means. However, it would be probably impossible to solve the riddle based on the RNA world hypothesis, because RNA with genetic information for protein synthesis never be generated in the absence of protein (Chap. 3, Sect. 4.8).

Then, the reasons are described in detail, why the origin of life remains unsolved still now, in the following Sect. 1, before the strategy for elucidating the enigma is explained in Chap. 2.

1 Why Has Not the Origin of Life Been Solved Still Now?

1.1 The Reasons from Viewpoint of Research Methodology

First, the reasons, why the riddle has been unsolved still now, are explained from viewpoint of research methodologies. Those are enumerated as follows.

1.1.1 The Research on the Origin of Life Usually Carried Out with the "Bottom-Up Approach"

Researchers working in the research fields of the origin of life would naturally try to understand what happened on the primitive Earth about 4 billion years ago, based on the so called "bottom-up approach". The bottom-up approach is of course quite important, because many events, which might occur on the primitive Earth, could be elucidated. However, the riddle of the origin of life could not be made clear only with the bottom-up approach, because it would be difficult to understand how even the core life system composed of three main members or gene, tRNA (genetic code) and protein was established (Ikehara 2016a). It might be also one of the reasons that many researchers have had a too deep interest in prebiotic synthesis of organic compounds as nucleotides on the primitive Earth.

1.1.2 Reproducibility Problem of the Events, Which Occurred on the Primitive Earth

Needless to state, "experiments" are quite important in the scientific research, because the fact can be confirmed through the experiments. However, the events, which might happen on the primitive Earth, could not be actually reproduced in a present laboratory, meaning that it is essentially impossible to confirm the events with experiments. Nevertheless, it would be of course indispensable to confirm whether or not an event supposed can be reproduced by experiments in a laboratory even in the life-origin research. For that reason too, it is important to experiment under a reliable theoretical consideration (see also Chap. 2: Sect. 1.3).

1.1.3 Long-Time Phenomena

It is well known that organic compounds were produced from inorganic compounds with prebiotic means during chemical evolution (for example, Miller and Orgel 1974). However, it might take thousands, even hundred thousands years or more until the organic compounds accumulated at a sufficient amount on the primitive Earth for the first life to emerge. Thus, the research on the origin of life must target such events, which might take place for a quite long time scale. This also makes it difficult to prove the events with experiments in a laboratory.

1.1.4 Repeating Phenomena

Extant organisms are living under the fundamental life system, in which gene leads protein synthesis. Therefore, a protein is always produced if the corresponding gene is expressed. A chemical reaction is always catalyzed by an enzyme, if the enzyme meets with the corresponding substrate. It is generally easy to confirm such a phenomenon with experiments in a laboratory.

On the contrary, many events progressing toward the emergence of life might be a pile of phenomena like that steps might proceed as accidental results among many phenomena repeatedly occurred. Therefore, it is supposed that many difficulties could be overcome to step forward by repeating the phenomena many times on the primitive Earth. In other words, it might be necessary for the emergence of life to repeat such many trials. It must be recognized that the study on the origin of life is to make clear such difficult events by experiments.

1.1.5 Another Difficulty in the Origin-Life Research

As described above, experiments are quite important even in the research on the origin of life. However, it would be impossible to unravel the origin of life, if experiments were carried out under a wrong idea. The correct answer could not be obtained from the experiments, which were carried out under such a wrong theory, even if the same experimental results were obtained reproduciblly in a present laboratory. This also makes it difficult to solve the riddle of the origin of life by the experiments.

1.2 How Could the Three Main Members in the Fundamental Life System Be Created Through Random Processes?

It is necessary to make clear at least how the three main members or gene, tRNA (genetic code) and protein were created on the primitive Earth in order to solve the riddle of the origin of life. However, the first three members must be created through random processes, because both structure and function of the first members never be naturally designed previously. However, both a gene encoding a mature protein and a mature protein like a precision polymer machine never be created directly through random processes, as described in Chap. 3: Sects. 4.2, 4.3, and 4.4 in detail. Note that tRNA also never be created at one stroke by random joining of nucleotides, because even a modern tRNA composed of about 75 nucleotides is too large to produce it through random process (see Chap. 6: Sect. 3.3). Solving the origin of life is to elucidate the difficult processes for establishing the fundamental life system.

1.3 Then, Is It Impossible to Unravel the Riddle of the Origin of Life?

Thus, the reason why the origin of life has been unsolved until now is because there exist many difficulties for solving the riddle. The origin of life is surely one of quite difficult problems, which it seems impossible to solve. Then, should we give up solving the riddle? On the other hand, if persons know a correct answer, which someone has solved, many persons would often consider that even the answer of such a quite difficult problem, is natural and it is easy to understand that. It is also an unquestionable fact that life arose on the primitive Earth about 4 billion years ago, as understood easily by seeing that we in fact are living on the present Earth, although, of course, I know that some researchers believe that life, which emerged on the other planets, arrived at the Earth and propagated itself on this planet. Anyhow those indicate clearly that the riddle of the origin of life must be elucidated someday.

2 My Idea About the Origin of Life

2.1 I Might Be Able to Solve the Riddle of the Origin of Life from My Own Viewpoint

I feel now that I could present reasonable steps to the emergence of life by verifying my ideas mainly with database analyses of base sequences of genes and amino acid sequences of proteins in extant microorganisms (Ikehara 2002, 2005). I could fortunately present a reasonable establishment process of the fundamental life system, including the respective establishment processes of the six members or protein, cell structure, metabolism, tRNA, genetic code and gene, according to [GADV]-protein world hypothesis or GADV hypothesis on the origin of life (Fig. 1.2a). My ideas, GADV hypothesis based on the origins of the six members, are described in short in this Introduction for readers to survey my researches and to understand how my researches progressed, before taking up the main subjects described in the respective Chapters of this book.

2.2 The Reason Why I Could Reach a Novel Idea, with Which the Riddle of the Origin of Life Might Be Solved

One more important matter about the reason, why I am in a situation I consider that I might solve the riddle of the origin of life, must be explained here. My research on the origin of life started in a laboratory of Department of Chemistry, Faculty of Science, Nara Women's University about 30 years ago from the study on the origin

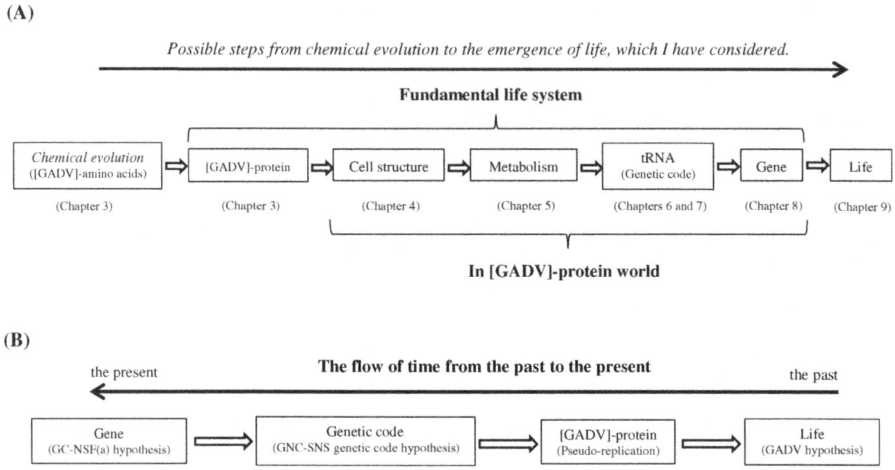

Fig. 1.2 **(a)** The probable steps from chemical evolution to the emergence of life according to [GADV]-protein world hypothesis or GADV hypothesis, which I have proposed. The reasons, why I have supposed the steps as shown in the figure, are described in detail in the respective Chapters of this book. It is assumed that life emerged from [GADV]-protein world, in which the fundamental life system composed of six members or protein, cell structure, metabolism, tRNA, genetic code and gene, was established. Eventually, life emerged after the establishment of the fundamental life system. **(b)** My research on the origin of life progressed from the study on how an entirely new gene has been created on the present Earth to GADV hypothesis on the origin of life, through GNC-SNS primitive genetic code hypothesis (Bold white arrows)

of modern genes, how entirely new (EntNew) genes have been created on the present Earth, which is apparently unrelated to the origin of life (Fig. 1.2b). Therefore, I did not at first intend to solve the riddle of the origin of life, but started from the study on the origin of modern genes, how and from where an EntNew gene has been created. Consequently, I fortunately reached one conclusion, "GC-NSF(a) hypothesis" on the origin of modern genes, suggesting that an EntNew gene has been created from a nonstop frame on antisense strand of a GC-rich gene (GC-NSF(a)) (Chap. 8; Sect. 3.3.4) (Ikehara et al. 1996). At that time, I calculated protein structural indexes with amino acid composition by accident (see Chap. 3; Sect. 4.4). The use of amino acid composition lead to one new concept, protein 0th-order structure, because all proteins, which were synthesized by random joining of amino acids in the same amino acid composition, have the same or at least similar amino acid composition and, therefore, have roughly the same protein structural index (see Chap. 3; Sect. 4.4). Successively, I could obtain several ideas on the origins of three members constituting the core life system, in order of GNC-SNS primitive genetic code hypothesis (Chap. 7, Sect. 3.1) (Ikehara et al. 2002), and protein 0th-order structure, which provides a scientific basis to pseudo-replication of [GADV]-protein, as described them in the following Chap. 3 in detail (Fig. 1.2b) (Ikehara 2009, 2014). After that, I noticed the idea, GADV hypothesis on the origin of life, assuming that life emerged from [GADV]-protein world, which was formed by random joining of [GADV]-amino acids in one of protein 0th-order structures or the amino acid composition or [GADV]-amino acids (Chap. 9) (Ikehara 2002, 2005). A rough outline of the steps to the emergence of life is given in Fig. 1.2a.

Therefore, I could fortunately and unintentionally meet the hypothesis on the origin of life, according to a top-down approach (Ikehara 2016a). Note that my top-down approach is quite different from the usual top-down approaches carried out by many researchers so far, for example, a search for coancestor with phylogenetic analysis (Woese 1987; Woese and Fox 1977) *etc*. Inversely considering this, it suggests that it would be difficult for many other researchers, who have explored based on the bottom-up approach and the usual top-down approaches, to meet the GADV hypothesis. Thus, as the results, I could fortunately reach the possible correct solution of the origin of life based on the establishment processes of the six members of the fundamental life system (Fig. 1.2).

I will discuss in the following Chapters whether or not the origins of the six respective members and the whole fundamental life system are valid, as considering the minimum conditions described in each Chapter, which are necessary to unravel the difficult creation processes of the six members and could lead solving the riddle of the origin of life.

I have now considered that the establishment process of the fundamental life system could be comprehensively explained under one concept "protein 0th-order structure" or "[GADV]-amino acids", which was obtained from the original viewpoint (Fig. 1.2a) (Ikehara 2014). I also consider that the reason, why I might solve the riddle of the origin of life, is because discovery of the "protein 0th-order structure", which may correspond to the discovery of "zero" in mathematics, led fortunately to the GADV hypothesis on the origin of life.

3 Overview of This Book

The GADV hypothesis is an idea, in which the origins of the six members can be explained comprehensively. The similar discussion may repetitively appear in the respective Chapters, because the six members, which are mutually and closely related through the protein 0th-order structure or [GADV]-amino acids in many cases, are entangled each other. Therefore, I have specified as showing what is described and discussed in the respective Chapters with the essentially same Fig. 1.2a at the first part of each Chapter as highlighted with a color box in the figure.

I further consider as follows. As well known, contemporary organisms inhabit under a complex system using diversified organic compounds. However, many researchers will agree with the idea that the first life would emerge under a much simpler system than the contemporary complex system. However, still many researchers may naturally suppose that at least tens of organic compounds should be contained even in the simple primitive system, although the first life could emerge easier, as a smaller number of kinds of organic compounds were used in the first life. I had also considered so at first. However, the GADV hypothesis, which I have reached, suggests that essentially one kind of polymer, [GADV]-protein, which was composed of [GADV]-amino acids, made an opportunity for the emergence of the first life, as shown in Fig. 1.2a.

Please read this book as understanding the reasons, which are described in detail in the respective Chapters, how and why I have considered on the origin of life. I am, of course, confident with my feeling that I have continued to explore the origin of life fairly, logically and reasonably thus far. For the purpose, I have intended to confirm with databases of genes, proteins and so on, whether or not my ideas are correct and reasonable.

However, many readers of this book might consider that I have piled up uncertain results or considerations based on the results, which were obtained under the self-centered investigations carried out selfishly. Certainly, I might unconsciously pile up unreasonable considerations on the considerations, which were obtained under self-centered and wrong ideas. So, please, feel free to indicate my misunderstandings and unreasonable points of the GADV hypothesis, as I may misunderstand many ideas especially for the details, which have been proposed by other researchers, as those are widely distributed. I have confidence in replying to every comment about the origin of life, which will be given by the readers, although, of course, I may answer some questions, as "I do not understand that now", if I do not really understand it as yet. However, I believe now that it would be probably impossible to give a correct answer to the same question according to other hypotheses. In addition, I will abandon the GADV hypothesis without hesitation, if critical mistakes, which I cannot answer appropriately, are indicated by readers, if more reasonable steps to the emergence of life were presented than the steps, which I have proposed, and if I judged that my idea is wrong, because what I want to know is only the mechanism under which life emerged on the primitive Earth and I do not want to hold fast my wrong idea. Inversely, I have felt that the steps to the emergence of life could not be reasonably explained by any other idea, which has been proposed by other researchers, and that, on the contrary, the reasonable steps to the emergence of life are presented by the GADV hypothesis.

Then, I will introduce the contents of this book briefly. I have described in order of the origins of protein (Chap. 3), cell structure (Chap. 4), metabolism (Chap. 5), tRNA (Chap. 6), genetic code (Chap. 7), gene (Chap. 8) and life (Chap. 9) (Fig. 1.2a). Note that the order of the description is reverse to the genetic flow and is along by the establishment processes of the respective members, which I consider. The reason is because the fundamental life system was established as going up the genetic flow from the most downstream of protein and also because gene never be created without [GADV]-protein, cell structure ([GADV]-microsphere), the initial metabolism and tRNA (anticodon stem-loop tRNA), as described in this book in detail (Chaps. 3, 4, 5 and 6).

I published one book entitled as "GADV hypothesis on the origin of life -Life emerged in this way!?" from LAP LAMBERT Academic Publishing about 4 years ago (Ikehara 2016b). I described in the book, the research process, through which I reached the GADV hypothesis, and the steps to the emergence of life, which have been supposed from the hypothesis. I, myself, consider that thereafter my research has largely progressed and I could solve many problems on the origin of life.

Therefore, I have described mainly interpretations about questions as "Why has the origin of life been unsolved still now?" and "What matters are necessary to solve the riddle of the origin of life?", which should lead solving the origin of life. The answers to the questions, which were obtained after publication of the book from LAP LAMBERT Academic Publishing, are described as comparing them with the hypotheses and theories, which have been proposed by other researchers.

References

Gilbert W (1986) The RNA world. Nature 319:618

Ikehara K (2002) Origins of gene, genetic code, protein and life: comprehensive view of life system from a GNC-SNS primitive genetic code hypothesis. J Biosci 27:165–186

Ikehara K (2005) Possible steps to the emergence of life: the [GADV]-protein world hypothesis. Chem Rec 5:107–118

Ikehara K (2009) Pseudo-replication of [GADV]-proteins and origin of life. Int J Mol Sci 10:1525–1537

Ikehara K (2014) Protein ordered sequences are formed by random joining of amino acids in protein 0th-order structure, followed by evolutionary process. Orig Life Evol Biosph 44:279–281

Ikehara K (2016a) Evolutionary steps in the emergence of life deduced from the bottom-up approach and GADV hypothesis (top-down approach). Life 6:6

Ikehara K (2016b) GADV hypothesis on the origin of life -life emerged in this way!? LAP LAMBERT Academic Publishing, Saarbrucken

Ikehara K, Amada F, Yoshida S et al (1996) A possible origin of newly-born bacterial genes: significance of GC-rich nonstop frame on antisense strand. Nucleic Acids Res 24:4249–4255

Ikehara K, Omori Y, Arai R et al (2002) A novel theory on the origin of the genetic code: a GNC-SNS hypothesis. J Mol Evol 54:530–538

Miller SL, Orgel LE (1974) The origins of life on the earth. Prentice-Hall, Englewood Cliffs

Woese CR (1987) Bacterial evolution. Microbiol Rev 51:221–271

Woese CR, Fox GE (1977) Phylogenetic structure of the prokaryotic domain: the primary kingdoms. Proc Natl Acad Sci U S A 74:5088–5090

Chapter 2
Modern Fundamental Life System

Abstract Objective of the study on the origin of life is to make clear how the fundamental life system composed of six members or gene, genetic code, tRNA, metabolism, cell structure and protein, was established. When we explore the origin of life, I first recommend to consider that organisms are living under the core life system composed of gene (DNA-mRNA; genetic information), tRNA (genetic code) and protein (Fig. 2.1a), rather than under the so called "Central dogma" (DNA, mRNA and protein) (Fig. 2.1b), because tRNA played important roles in establishing the fundamental life system as explained in Chap. 8: Sect. 3.1. On the other hand, it is important to confirm whether or not the first three members could be formed through random processes, only which occurred on the primitive Earth. For the purpose, general features of the six modern members are described in every Chapter in order of protein, protocell, metabolism, tRNA, genetic code and gene or in order of the members which would be formed until the first life emerged, because it is necessary to understand well the features of the respective members to confirm whether or not the most primitive members assumed are suitable to the most primitive forms leading to modern ones. In addition, it is also important to confirm whether or not the most primitive members produced with the random process could evolve to the respective modern ones. Therefore, matters necessary to unravel the riddle of the origin of life are described in this Chapter. Outline of the establishment process of the fundamental life system or the steps from chemical evolution to the emergence of life is also described.

Keywords Origin of life · Fundamental life system · Core life system · Central dogma · [GADV]-protein world hypothesis (GADV hypothesis)

K. Ikehara, *Towards Revealing the Origin of Life*,
https://doi.org/10.1007/978-3-030-71087-3_2

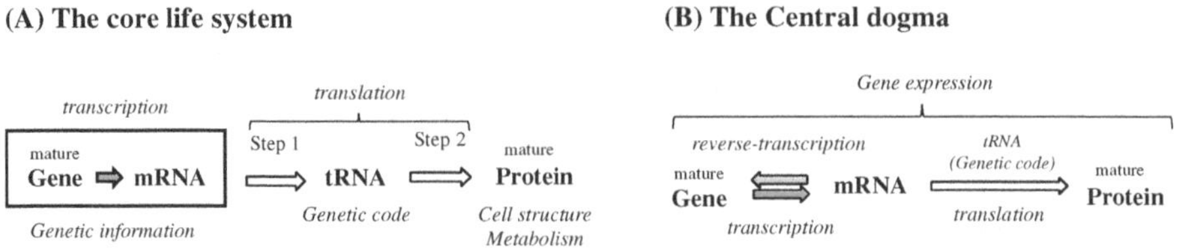

Fig. 2.1 (**a**) The core life system composed of gene, tRNA (genetic code) and protein (cell structure, metabolism). I will discuss the origin of life mainly based on not the Central dogma but the core life system in this book. (**b**) The "Central dogma", which was proposed by Crick: A mature protein is produced through gene expression or transcription and translation followed by protein structure formation (Crick 1970). Base sequence of mRNA can be reverse-transcribed as shown by a brown bold left arrow

1 Towards Solving the Riddle of the Origin of Life

1.1 The Mechanism How Life Is Living

The most important point for elucidation of the origin of life is of course to clarify the establishment processes of the six members, or protein, cell structure, metabolism, tRNA, genetic code and gene, composing the fundamental life system, under which contemporary organisms are also living. Therefore, one of the premises for solving the origin of life is to well understand features of the six members of modern organisms, because even the most primitive members should be connected with the respective modern ones.

It has been considered thus far that diverse and splendid organisms, which are inhabiting on the present Earth, are living under the life system called as the "Central dogma" (Fig. 2.1b). Certainly, it seems that all extant organisms are living in a "gene-centered world" as considered by the Central dogma, because genes determine amino acid sequences of all proteins under the system and because a base substitution results in an amino acid replacement of a protein encoded by the gene at a high probability. On the other hand, Francis Crick pointed out the key point about the central dogma of molecular biology in 1970 as follows (Crick 1970).

> *The central dogma of molecular biology deals with the detailed residue-by-residue transfer of sequential information. It states that such information cannot be transferred from protein to either protein or nucleic acid.*

1.1.1 Problem About the "Central Dogma"

The description by Crick is, of course, correct, because the amino acid sequence information of a protein cannot be transferred to both base sequence of a gene and another amino acid sequence under the genetic system, although the base sequence information can be easily transferred between DNA and RNA (Fig. 2.1b). The Crick's insight, which has clearly instructed us about the impossible information transfer from a protein to a gene or another protein, was excellent. However, this might cause to generate a gene-centered idea, which instead has made it difficult to solve the origin of life. The reason is as follows.

The role of mRNA is overestimated in the Central dogma, because mRNA carries simply a genetic information received from DNA (Fig. 2.1b). It is also well known that base sequence of RNA can be reverse-transcribed into DNA (Fig. 2.1b), meaning that RNA shares the same genetic information with DNA. On the contrary, it is obvious that transcription is quite different from translation process. Therefore, the system should be redrawn as combining the two, gene and mRNA, both of which have essentially the same function as genetic information carriers and instead as highlighting tRNA as shown in Fig. 2.1a.

1.1.2 Proposition of "Core Life System" Instead of the "Central Dogma"

As well known, tRNA or genetic code plays an important and skillful role in mediating codon sequence of a gene with amino acid sequence of a protein in the genetic system. This can be understood from that it would be impossible to reverse-translate an amino acid sequence of a protein to a base sequence as indicated by Crick (1970) because of the absence of one-to-one correspondence between an amino acid and a nucleobase. Therefore, the roles of tRNA and genetic code should be put in a more prominent position in the genetic system. Inversely stating this, tRNA and genetic code has been underestimated thus far in the system, which was named by Crick as the "Central dogma" (Fig. 2.1b) (Crick 1970). Reevaluation of the significance of tRNA or genetic code in the genetic system is quite important, especially when the establishment process of the life system is explored, because the most primitive gene determining an amino acid sequence of the first mature protein would never be created independently of tRNA, and because the genetic sequence must be formed by connecting anticodons of the most primitive anticodon stem-loop (AntiC-SL) tRNAs as referring protein structure formation (see Chap. 6: Fig. 6.12) (Ikehara 2019). Therefore, I would like to point out here that extant organisms are living under the core life system composed of three members, gene (DNA-mRNA; genetic information), tRNA (genetic code) and protein (Fig. 2.1a). Thereafter, the term "core life system" is used to show the genetic system instead of the "central dogma" in this book, unless especially specified. Terms, the core life system and the fundamental life system, are used, when the main three members or gene, tRNA and protein and the six members or the three members plus cell structure, metabolism and genetic code, are expressed, respectively.

1.2 Necessary Matters for Solving the Riddle of the Origin of Life

Life or organisms are living under the fundamental life system composed of the six members or protein, cell structure, metabolism, tRNA, genetic code and gene. The six members used in modern organisms have been optimized to form the respective splendid structures and functions.

1. Therefore, it is necessary to understand well the features of the six respective modern members, in order to elucidate from what primitive ones the respective contemporary members originated.
2. It is also necessary to make clear formation processes of the most primitive six members through random processes in order to solve the riddle of the origin of life.
3. In addition, the origins of the respective six members must be interpreted comprehensively, because the six members should be mutually and deeply interrelated. In parallel, the formation processes of all the mos primitive six members should be explained under the essentially same concept. Inversely stating this, it should be highly likely that many contradictions arise among the ideas, which were obtained by independent explorations of the six members.
4. Furthermore, it is also important to explain how the six primitive members, which were supposedly created during random processes, evolved to the respective modern ones.
5. Finally, it must be clarified how the whole fundamental life system could be established

1.3 Significance of Theoretical Approach

1.3.1 The Origin of a Member Must Be Narrowed Down to a Reasonable One with Theoretical Approach

Generally, it would be difficult to prove the events, which happened on the primitive Earth about 4 billion years ago, with experiments in a laboratory (Fig. 2.2a) (Chap. 1: Sect. 1.1). Therefore, it is important to narrow down the events to ones which would occur at a high probability as much as possible, with theoretical approaches (Fig. 2.2b), because it is difficult to prove the events only with experiments. In such a case, analysis with databases of base sequences of genes and amino acid sequences of proteins would be one of the powerful approaches, because the databases have been obtained by experiments and, therefore, the results obtained by the database analysis should be equivalent to the experimental results.

I consider also that one of the reasons, why we can investigate on the origin of life, is because we, who are living in the contemporary age, can use and analyze databases of base sequences of genes and tRNAs, amino acid sequences of proteins

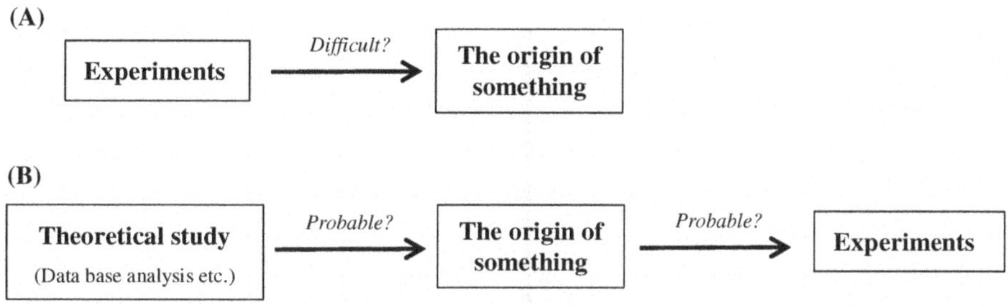

Fig. 2.2 (**a**) In the study on the origin of life, the events, which it is difficult to prove only with experiments, must be explored. (**b**) Therefore, it is important to confirm previously and theoretically whether or not there are possible steps to reach to the emergence of life. In such a case, it would be effective to narrow down into the events, which would occur on the primitive Earth at a high probability as much as possible, by theoretical approaches, for example as using databases of base sequences of genes and amino acid sequences of proteins and after that to confirm them with experiments

in addition to three-dimensional structures of proteins and tRNAs, which are stocked in various databanks.

1.3.2 Even the Most Primitive Member Must Have Basic Features Similar to Modern One

As described so far, the important matter for solving the origin of life should be to understand the ways how all of the six primitive members were created through the respective random processes, either directly or indirectly. That is,

1. It is important to clarify creation processes of the most primitive members having similar features to the modern ones through random processes.
2. Features of some members assumed, which were formed before the first life emerged, might be dissimilar to the basic features of the modern members (Fig. 2.3a). In such cases, as a matter of course, the primitive member with different features from modern one must be first converted to the most primitive one with the basic features similar to the modern member (Fig. 2.3(A) (2)). Therefore, the features of modern members can be effectively used to judge whether the primitive members assumed are suitable to the original ones or not. Validity of the primitive one assumed can be judged by whether transition process of the members from "have not" to "have" the basic features can or cannot be reasonably explained irrespective of directly or indirectly.

I will explain about a few examples in more detail below that it could not be concluded that the modern member originated from something assumed as the most primitive one.

(1) It is assumed in the RNA world hypothesis that modern proteinaceous catalysts originated from catalytic RNAs or ribozymes and, therefore, the catalytic activity on the RNA was transferred to a protein (Gilbert 1986). In the case, the

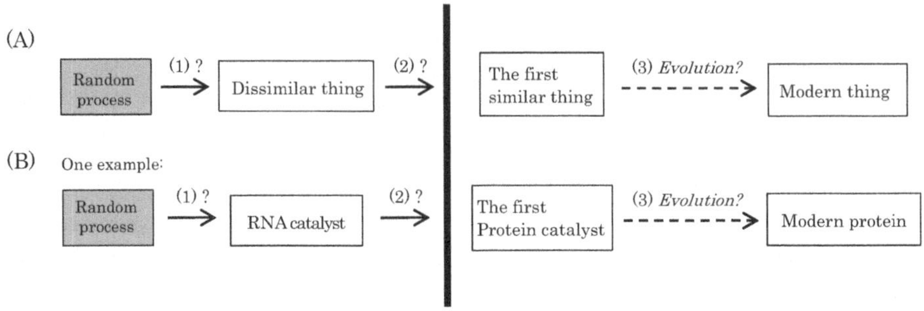

Fig. 2.3 (**a**) Consider a case that feature of a primitive member assumed was dissimilar to the feature of the modern one. It should be judged that the dissimilar thing assumed is not the origin of the member, if the following three points could not be satisfied. (step 1) The primitive member with different feature from modern one must be produced through random process. (step 2) The dissimilar thing must be first transformed to the most primitive one having the similar feature to modern one. (step 3) Next, the first similar thing must evolve to the modern member. Therefore, if the evolutionary process from the dissimilar thing to modern one cannot be reasonably explained, the primitive member assumed should be judged that it is not the original one of the modern one. (**b**) In RNA world hypothesis, it is assumed that protein catalysts (enzymes) originated from RNA catalysts (ribozymes) (Gilbert 1986). It must be judged that RNA catalysts were not the original ones of modern protein enzymes, (step 1) if the RNA catalysts could not be produced through a random process, (step 2) if the transition process from RNA catalysts to the most primitive protein catalysts could be unexplained reasonably either directly or indirectly, and also (step 3) if the evolutionary pathway from the most primitive protein catalysts to modern proteins could be unexplained

RNA catalysts could not be admitted to be the origin of protein enzymes, if it was impossible to transfer the catalytic activity on the RNA to protein either directly or indirectly (Fig. 2.3(B) (2)).

(2) Surface metabolism theory, which was based on pyrite, has been proposed to explain the origin of metabolism (Wächtershäuser 1990). In the case too, catalytic activity on the surface of inorganic mineral or pyrite must be transferred to a protein one day, if the surface metabolism theory is correct. However, it would be impossible to transfer the catalytic activity on the surface of pyrite to a protein (Chap. 5: Sect. 2.1). Therefore, it cannot be admitted that the catalytic reactions on pyrite supposed in the surface metabolism theory were the origin of modern metabolism, which is driven by protein enzyme.

(3) Protein-origin theory based on a self-replicating protein, amyloid, (Maury 2009), has been proposed to explain the origin of life. However, in this case too, it would be impossible to transfer self-replication ability of the amyloid to double-stranded RNA or DNA (see also Chap. 9, Sect. 2.3.1). Therefore, the idea advocating that RNA or DNA replicator originated from the amyloid, which could be self-replicated, cannot be accepted as correct one for the origin of life, because the replication ability of the amyloid could not be transferred to RNA or DNA.

1.3.3 Something Assumed as the Origin Must Be the Most Primitive One, Which Could Evolve to Modern One

It is also important to confirm with theoretical approaches in advance, whether or not something assumed as the origin is the simplest or the most primitive one and there are possible steps to reach from the something to the emergence of life (Fig. 2.3a (3)), because, there are some cases, assuming that the something might be too strongly affected by modern one to grasp correctly the most primitive one. For example, a part of the whole structure of modern one has been frequently assumed to be the original one. In such cases, it would be important to satisfy the following conditions, because it would be difficult to prove the matters though experiments. These would also indicate the significance of theoretical approaches for the studies on the origin of life.

(1) Something assumed must be the simplest one, which can evolve to the modern one. In other words, it must be confirmed that there does not exist a simpler one than the something assumed.
(2) It must be confirmed that the something assumed could be produced through random process on the primitive Earth.
(3) It must be also confirmed that the whole structure could be constructed from the part structures, which were assumed to be the original one.

I will explain about a few examples showing significance of theoretical approaches.

(1) A minihelix composed of T-stem loop and accepter stem, which are two parts of modern L-form tRNA, has been assumed as the original one of modern tRNA. However, it would be quite difficult to create the minihelix by random joining of nucleotides and also for the minihelix to evolve to L-form tRNA (Tamura 2015)) (see Chap. 6: Sect. 2.1). In addition, anticodon stem-loop (AntiC-SL) of L-form tRNA is more simple than the minihelix. Therefore, it should be judged that the AntiC-SL is more appropriate for the origin of modern tRNA than the minihelix, if it is explained that the AntiC-SL could be formed through random process and could evolve to modern L-form tRNA.
(2) Short peptide (motif) theory has been proposed to explain how modern proteins could be produced (Dayhoff 1965; Romero et al. 2016). This theory is a protein version of exon-shuffling theory on the origin of gene (Chap. 8: Sect. 2.2). In this case too, it would be quite difficult to generalize the construction strategy of a modern protein by assembly of short peptides or motifs, because it is considered that a modern protein has been generally formed not by assembly of short peptides or motifs of proteins (Chap. 3: Sect. 3.3), but through evolutionary process from an immature protein to a mature protein (Chap. 3; Sects. 4.2, 4.3, and 4.4).

On the other hand, all the six modern members have the respective specific and amazing structures and functions because of the evolution of those during about 4 billion years since the first life emerged. All the six modern members never be produced directly through random process based upon commonsense. However, only random reactions naturally proceeded on the primitive Earth. Therefore, elucidating the origins of the respective members is to understand the way how such a splendid modern member could be produced from the most primitive one, which satisfies at least the two minimum conditions described below (Table 2.1). The objective of this book is to aim at exploration of the origin of life as comparing the establishment processes of the fundamental life system based on GADV hypothesis, which I have proposed, and other hypotheses.

In fact, I have advanced the research on the origin of life based on the matters, which I have presumed as correct from the results obtained by data analyses (Ikehara 2002, 2005). From the results, I believe now that I could obtain one reasonable evolutionary pathway from chemical evolution to the emergence of life, which is quite different from other hypotheses (Fig. 2.4).

Table 2.1 Two minimum conditions, which the most primitive members of the fundamental life system must satisfy

The six members	1st condition	2nd condition
Protein	Water-soluble globular structure	Expression of catalytic activity
Metabolism	Catalytic reactions driven by protein enzyme	Catalyzing organic compounds accumulated
Cell structure	Cell membrane	Transport of chemical compounds
tRNA	Mediating between codon and amino acid	Plural numbers and kinds of tRNAs
Genetic code	Correspondence between codon and amino acid	Determined by tRNA
Gene	Codon sequence	Double-stranded structure

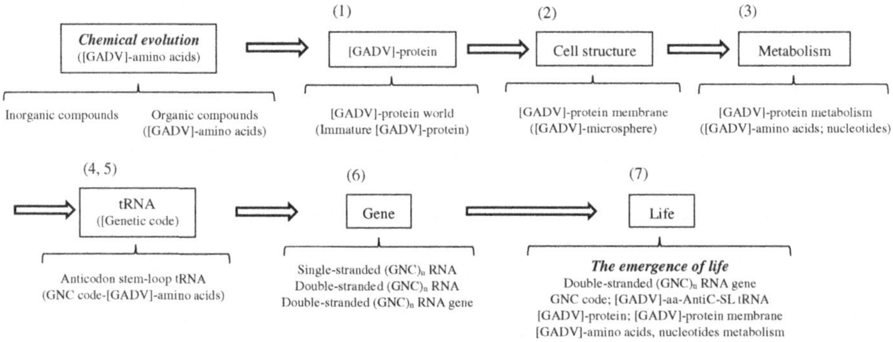

Fig. 2.4 Evolutionary steps from chemical evolution to the emergence of life. All formation processes of the six members composing the fundamental life system are related to [GADV]-amino acids and/or [GADV]-protein, as chemical evolution: [GADV]-amino acids produced by prebiotic means, (1) [GADV]-protein: [GADV]-protein world composed of immature [GADV]-proteins, (2) Cell structure: [GADV]-microsphere (protocell) surrounded by [GADV]-protein membrane, (3) Metabolism: Primitive metabolic system driven by immature [GADV]-proteins, (4) tRNA: AntiC-SL tRNAs bridging a [GADV]-amino acid with a GNC codon, (5) Genetic code: GNC genetic code encoding [GADV]-amino acids, (6) Gene: $(GNC)_n$ gene encoding [GADV]-protein, and (7) Life: The first genuine life emerged upon establishment of the fundamental life system composed of the six members

2 Overview of [GADV]-Protein World Hypothesis (GADV Hypothesis)

2.1 Establishment Process of the Fundamental Life System (Outline)

I will describe only a summary of the process, how the fundamental life system composed of gene, tRNA (genetic code), cell structure and protein (metabolism), was established, for the readers to catch the outline of this book and to understand developmental steps from organic synthesis with prebiotic means to the emergence of life (Fig. 2.4), as I will describe the respective formation processes after Chap. 3 in detail,

(1) Immature but pluripotent [GADV]-protein could be produced by random joining of [GADV]-amino acids in protein 0th-order structure (Chap. 3).
(2) Protocell surrounded by [GADV]-protein membrane ([GADV]-microsphere) was formed. In the protocell, a primitive metabolic system was prepared and the microsphere without any genetic material proliferated, accompanied by [GADV]-protein synthesis with immature [GADV]-protein or pseudo-replication of the [GADV]-protein, which caused a high osmotic pressure in the microsphere. Microspheres, which could proliferate at a higher rate than others, were selected at a higher probability (Chap. 4).
(3) [GADV]-amino acids, sugars including ribose, nucleotides and oligonucleotides were produced under the primitive metabolic system in [GADV]-microspheres or in [GADV]-protein world (Chap. 5).
(4) Successively, the most primitive AntiC-SL tRNA was created (Chap. 6).
(5) The first GNC code was established frozen-accidentally with four nonspecific AntiC-SL tRNAs (Chap. 7).
(6) Further, a single-stranded (ss)-$(GNC)_n$ RNA was formed by random joining of GNC anticodons carried by AntiC-SL tRNAs and double-stranded (ds)-$(GNC)_n$ RNA was produced by complementary strand synthesis of the ss-RNA (Chap. 8). Creation of the ds-$(GNC)_n$ RNA made it possible to evolve an immature [GADV]-protein to a mature one owing to the ds-RNA, which can memorize amino acid replacements of the evolving protein (Chap. 8).
(7) Finally, the first life emerged on the primitive Earth about 4 billion years ago (Chap. 9).

2.2 The Steps from Chemical Evolution to the Emergence of Life

As described above, elucidation of the establishment process of the fundamental life system composed of gene, tRNA (genetic code), cell structure (metabolism) and protein is indispensable to make clear the origin of life. Nevertheless, it seems to me that even the researchers studying on the origin of life have been almost careless about the key for solving the riddle of the origin of life, "protein 0th-order structure", or have considered without recognizing the key for solving the origin of life. Therefore, they must face up the difficult problem as long as the researches studying on the origin of life are carried out from the standpoint of the "gene-early theory".

On the contrary, I believe that I could solve the riddle of the origin of life. The reason is because I consider that the establishment processes of all of the six constituents could be explained reasonably in order of protein, protocell, metabolism, tRNA, genetic code and gene, which are necessary to solve the origin of life (Fig. 2.4). I will describe and discuss the establishment processes of the six members in the following Chapters of this book in detail, which lead to [GADV]-protein world hypothesis or GADV hypothesis.

The order of evolutionary steps from chemical evolution to the emergence of life should be almost fixed, because an event happened at one step before caused the event happened after that. For examples, the most primitive anticodon stem-loop tRNA (AntiC-SL tRNA) could be produced because of the initial metabolism, which was driven by immature [GADV]-proteins. The first $(GNC)_n$ gene was created by repeatedly connecting two anticodons of two AntiC-SL tRNAs, which were randomly juxtaposed. This means that AntiC-SL tRNA must be created after a initial metabolism was formed and before the first gene was created so that the steps to the emergence of life can be reasonably explained.

References

Crick F (1970) Central dogma of molecular biology. Nature 227:561–563

Dayhoff MO (1965) Computer aids to protein sequence determination. J Theor Biol 8:97–112

Gilbert W (1986) The RNA world. Nature 319:618

Ikehara K (2002) Origins of gene, genetic code, protein and life: comprehensive view of life system from a GNC-SNS primitive genetic code hypothesis. J Biosci 27:165–186

Ikehara K (2005) Possible steps to the emergence of life: the [GADV]-protein world hypothesis. Chem Rec 5:107–118

Ikehara K (2019) The origin of tRNA deduced from *Pseudomonas aeruginosa 5′* anticodon-stem sequence: anticodon stemloop hypothesis. Orig Life Evol Biosph 49:61–75

Maury CPJ (2009) Self-propagating β-sheet polypeptide structures as prebiotic informational molecular entities: the amyloid world. Orig Life Evol Biosph 39:141–150

Romero MLR, Rabin A, Tawfik DS (2016) Functional proteins from short peptides: Dayhoff's hypothesis turns 50. Angew Chem Int Ed 55:15966–15971

Tamura K (2015) Origins and early evolution of the tRNA molecule. Life (Basel) 5:1687–1699

Wächtershäuser G (1990) Evolution of the first metabolic cycles. Proc Natl Acad Sci U S A 87:200–204

Chapter 3
The Origin of Protein

Abstract Every contemporary protein or a mature protein having a splendid structure and function is produced through genetic expression under the present genetic system. On the other hand, the ideas, such as the amino acid sequence hypothesis (Crick, FH, Symp Soc Exp Biol 12:138–163, 1958; Dill KA, Biochemistry 29:7133–7155, 1990), the protein structure hypothesis (Dill KA, Biochemistry 29:7133–7155, 1990), the short peptide (or motif) hypothesis (Dayhoff MO, J Theor Biol 8:97–112, 1965; Romero K, Amada F, Yoshida S et al., Nucl Acids Res 24:4249–4255, 2016) and so on, have been proposed to explain the origin of protein. However, any hypothesis could not explain well the creation process of the first primitive protein, probably because those three hypotheses are too strongly affected by formation process and structure of modern proteins. Then, how did modern proteins originate from what structure of a primitive protein? As a matter of course, the most primitive protein must be created through random process. On the other hand, it would be certain that it is impossible to produce any mature protein in the absence of gene or through a random process, because of the high wall of the extraordinary large diversity of amino acid sequences composed of 100 amino acids; $20^{100} = \sim 10^{130}$ (Dill 1990). However, not a mature protein but an immature protein with some flexibility could be produced by random joining of [GADV]-amino acids in a specific amino acid composition or protein 0th-order structure, as I have proposed. That is, the mature protein had been created as a result of maturation from an immature protein to a mature protein with a higher function than before step by step, as enhancing a function of the immature protein with memorizing ability of amino acid replacements of double-stranded (ds) RNA. Note that the ds-RNA never play a leading role in creation of the mature protein during the maturation process, because the random base substitutions of the ds-RNA cannot determine the evolutionary process of the immature protein. Instead, the evolutionary pathway of the protein is actually determined by improvement of the function of the evolving protein itself. Therefore, the ds-RNA only had played a subsidiary role during the maturation from an immature to a mature protein. I will discuss in this Chapter, how such a splendid mature protein could be produced from an immature protein, dividing into two parts, or before and after creation of the first ds-$(GNC)_n$ RNA gene.

K. Ikehara, *Towards Revealing the Origin of Life*,
https://doi.org/10.1007/978-3-030-71087-3_3

Keywords Origin of protein · Amino acid sequence hypothesis · Protein structure hypothesis · Short peptide (motif) theory · Homochirality · [GADV]-amino acids · Protein 0th-order structure · Immature pluripotent protein

In an introductory article of Daniel E. Koshland, Jr., "The key-lock theory and the induced fit theory", it is described (Koshland 1995), as follows.

> *As a glove changes shape when a hand slips into it, so an enzyme changes its conformation on binding a ligand. This theory of induced fit extends the lock-and-key principle that Emil Fischer proposed exactly 100 years ago. The new theory proposed by D. E. Koshland, Jr. in 1958 allows one to explain regulation and cooperative effects, and adds some new specificity principles as well.*

As proposed by Koshland, it is well known that an enzyme changes its conformation upon binding with a ligand (Koshland 1995). It can be well understood from a three-dimensional structure of a protein as an aminoacyl-tRNA synthetase shown in Fig. 3.1 that every modern protein is constructed so beautifully. Nevertheless, even such a highly qualified protein can be easily produced under the genetic system. On the other hand, as a matter of course, the first primitive protein should be produced through a random process, only which proceeded on the primitive Earth, because structure and function of the protein could not be designed previously. However, no mature protein could be created by random joining of amino acids at one stroke. Then, how was the amazing protein created through random process especially on the primitive Earth? The purpose of this Chapter, in which the origin of protein is discussed, is to consider the way how such a primitive protein could be created through random process in the absence of gene (Fig. 3.2).

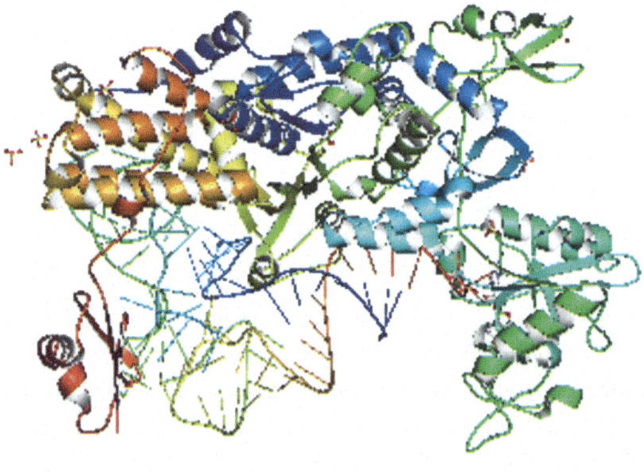

Fig. 3.1 Three dimensional structure of a complex of Leu-tRNA and Leu-tRNA synthetase (Leu-ARS) (PDBj; 2V0G). Leu-ARS is bound with Leu-tRNA at mainly three points, anticodon stem-loop, accepter stem and $^{5'}$CCA$^{3'}$-end of tRNA as holding the tRNA with the respective domains. How could such a splendid ARS be created? It would be impossible to generate the ARS with random process at one stroke. Therefore, it is necessary to create ARS through maturation process from an immature ARS to the mature ARS. Interpretation of the origin of protein might assist to clarify the creation process of protein like as ARS

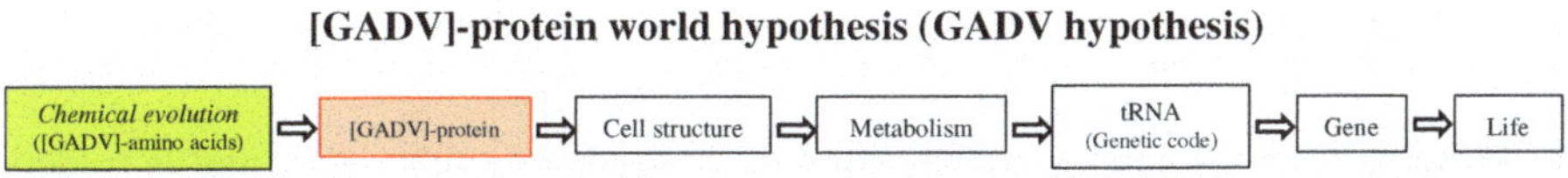

Fig. 3.2 In this Chapter, the origin of protein is discussed (red box). Synthesis of the most primitive protein ([GADV]-protein) was supported by supply of [GADV]-amino acids, which were produced with prebiotic means during chemical evolution (yellow box). Note that all the six members are connected in a row as shown with white bold arrows

1 Introduction: Towards Solving the Riddle of the Origin of Protein

First, it is described what matters are necessary to explore and solve the riddle of the origin of protein. It is discussed as dividing the general features of modern proteins into structural and functional features, because it can be checked from the two sides whether or not something assumed as the most primitive one is suitable to the origin of protein.

1.1 Necessary Matters for Exploring the Origin of Protein

1. It is necessary to well understand what features modern proteins have, in order to clarify the origin of protein, a member of the fundamental life system. The reason is because it is indispensable to confirm whether a primitive one, which is assumed in a hypothesis, could be created by random reactions on the primitive Earth and the primitive one assumed could evolve to the contemporary protein or not (Chap. 2: Fig. 2.3a). In other words, it must be made clear from what kind of primitive one the contemporary protein originated and how the primitive one evolved to modern protein.
2. If it is assumed that contemporary proteins originated from a primitive something with different features of modern proteins, as a matter of course, the transition process from the something to the modern protein must be reasonably explained, either directly or indirectly (Chap. 2: Fig. 2.3a). The reason, why I take trouble to explain this here, is because in some hypotheses, something (for example, RNA catalyst in the RNA world hypothesis) substantially different from modern one (protein catalyst) is actually assumed. In such cases, it becomes important whether catalytic activity of RNA different from modern protein can be transferred to protein.

Then, before exploration of the origin of protein, general features of modern proteins, above all, of the representative protein, enzyme, are first enumerated, as dividing into two parts of structural and functional features.

1.2 General Structural Features of Modern Protein (Enzyme)

1. Protein, which is generally composed of 20 kinds of amino acids and is folded into three-dimensional structure, has amazing structure and function as like a precision polymer machine.
2. Protein generally contains secondary structures as α-helix and β-sheet. Inversely stating this, why protein does not exist, which is composed of only amino acids (Pro, Gly, Ser, Asn and Asp) having a propensity to turn/coil structure? The reason is probably because such a protein, which is composed of only turn/coil forming amino acids, could not be folded into a stable water-soluble globular structure.
3. Protein contains both hydrophobic amino acids and hydrophilic amino acids in appropriate proportions so that the polypeptide chain can be folded into a globular structure with a rather high rigidity in water. The rigidity makes it possible to accept one specific substrate at a catalytic center of the enzyme and to reject any other organic compounds from the site, like as a key and a key hole (Koshland 1995).

1.3 General Functional Features of Modern Protein (Enzyme)

1. Protein enzyme generally exhibits only one catalytic activity.
2. All organisms are living with proteins composed of only 20 amino acids on this planet. This clearly indicates that all catalytic functions necessary for all organisms to live on this planet, can be produced essentially with only 20 natural amino acids, although it is known that some amino acids are chemically modified post-translationally.

Exploring the origin of protein is to explain the way how a primitive something having or not having the fundamental features described above (1) could be produced through random processes and (2) could evolve to a modern protein (Sect. 1.1). Then, it is first examined whether or not hypotheses, which have been proposed by other researchers, are valid from the viewpoints of the two necessary conditions for the origin of protein.

2 Structure Formation of Protein Under the Modern Genetic System

As well known, every modern protein like a precision polymer machine is formed after the expression of a genetic function (Fig. 3.3). That is, (1) a polypeptide chain with an amino acid sequence (primary structure) is produced through expression of

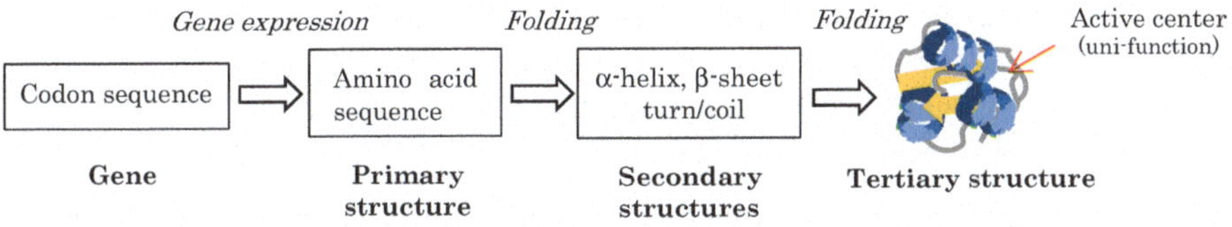

Fig. 3.3 Gene expression and formation of a mature protein. Under the modern genetic system, a polypeptide chain with an amino acid sequence (primary structure) is synthesized by expression of a codon sequence of a gene followed by translation of the codon sequence. The polypeptide chain is folded into the respective secondary structures to form a tertiary structure through assembling them to a hydrophobic core. Thus, contemporary and mature proteins like a precision polymer machine are always produced under the genetic function. Note that such a splendid protein never be produce through random process in the absence of the corresponding gene

a gene. (2) The polypeptide chain is folded into secondary structures, as α-helix, β-sheet and turn/coil conformations, depending on amino acid sequence of the respective peptide segments. (3) The segments folded into secondary structures are generally assembled to a hydrophobic core to form a rather rigid tertiary structure. A completed protein or an enzyme generally has one catalytic center binding with one substrate through specific interaction.

Understanding the formation process of an entirely new (EntNew) gene/protein should be one of the most important issues in the field of biological sciences. Nevertheless, it is totally unknown still now, how a mature protein encoded by a gene has been created. In every textbook of "Biochemistry" or "Molecular biology" (for example, Stryer's "Biochemistry"; Berg et al. 2002), formation of a mature water-soluble globular protein always starts from an amino acid sequence of a protein or a protein primary structure, which is specified by gene expression, or transcription and translation of a gene (Fig. 3.3). Therefore, it may seem that the key for production of a protein is always held by the genetic sequence encoding one dimensional amino acid sequence (Fig. 3.3). Certainly, all modern proteins are produced under the gene expression and a catalytic activity will be changed by a base substitution of the genetic sequence at a high probability. Thus, it can be concluded that gene always controls protein synthesis under the modern genetic system.

3 The Ideas on the Origin of Protein Advocated by Other Researchers

Two sides should be considered for exploration of the origin of protein. The first one is to explore the mechanism how the first mature protein was created (the origin of protein-I). This is the intrinsic origin of protein. The second one is how EntNew proteins, which have not meaningful homology with any other previously existed proteins, were and have been created if necessary (the origin of protein-II), even after the first protein was created.

3.A The Origin of Protein-I: How Was the First Mature Protein Created?
However, I do not know the ideas on the origin of protein-I, which have been studied from the first side and have been proposed by other researchers in order to answer the question, "How was the first mature protein created?"

3.B The Origin of Protein-II: How Have Entirely New Mature Proteins Been Created?
Gene always controls protein synthesis. Therefore, amino acid sequence hypothesis, in which the ground of the origin of protein-II is sought to an amino acid sequence, has been proposed as a natural idea (Crick 1958; Dill 1990). Then, start the discussion of the origin of protein-II from the amino acid sequence hypothesis.

3.1 Amino Acid Sequence Hypothesis

In the amino acid sequence hypothesis, it is considered that a unique sequence must be first synthesized to produce a globular protein carrying a specified function. The idea is correct and is accepted in this regard by many persons, because the amino acid sequence hypothesis is a natural idea, if the unique amino acid sequence for the mature protein could be created in some way. However, as well known, even a small protein composed of 100 amino acids never be produced by random joining of amino acids, because the sequence diversity reaches an extraordinary large value as $20^{100} = \sim 10^{130}$, since modern proteins are composed of 20 kinds of amino acids (Fig. 3.4) (Dill 1990). Therefore, it is actually impossible to produce any EntNew protein by random joining of amino acids.

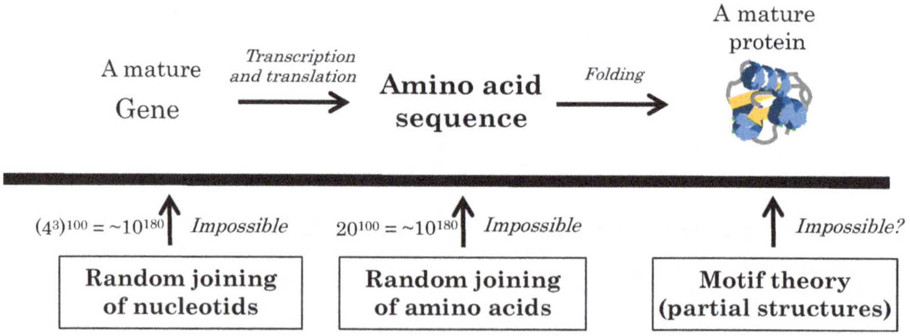

Fig. 3.4 It would be impossible to create a mature protein by random joining of amino acids because of the extraordinary large amino acid sequence diversity of $\sim 10^{130}$ (Dill 1990). Similarly, it would be also impossible to create a gene encoding a mature protein through random process because of the extraordinary large nucleotide sequence diversity of $(4^3)^{100} = \sim 10^{180}$. In addition, a mature protein could not be created by assembling partial structures or motifs (Sect. 3.3). Therefore, it is unfortunately impossible to solve the problem on the origin of protein-II through random process

The one way generating an unique amino acid sequence is to synthesize the amino acid sequence under the genetic function. This means that the problem on the EntNew protein creation can be solved, if a gene encoding the amino acid sequence of a protein could be produced by random joining of nucleotides, or if the way can be understood, how genetic sequence encoding the amino acid sequence could be generated under random process on the primitive Earth. However, it is also obvious that any gene encoding a mature protein cannot be created through random joining of nucleotides, because the diversity of nucleotide sequence is also extraordinary large as $(4^3)^{100} = \sim 10^{180}$ (Fig. 3.4). Therefore, any polynucleotide, which was produced by random joining of nucleotides, would be a simple chemical substance without any genetic information for protein synthesis, and a polypeptide, which was synthesized with the polynucleotide carrying a random nucleotide sequence, should be also a simple chemical substance without any function (Fig. 3.5).

Strengths of amino acid sequence hypothesis: The amino acid sequence hypothesis is a natural idea, because formation of a protein with tertiary structure always starts with an amino acid sequence produced under the modern genetic system as was rightly pointed out by Crick (Fig. 3.3 and Chap. 2; Fig. 2.1a) (Crick 1958, 1970).

Weaknesses of amino acid sequence hypothesis: Therefore, the problem on the origin of protein-II could be solved, if an amino acid sequence for the mature protein could be created directly by random joining of amino acids (Fig. 3.4) or indirectly through a gene encoding the protein, which was produced by random joining of nucleotidess (Fig. 3.4). However, the two routes are closed as described above (Fig. 3.4). Therefore, it means that the origin of protein-II cannot be solved with the amino acid sequence hypothesis.

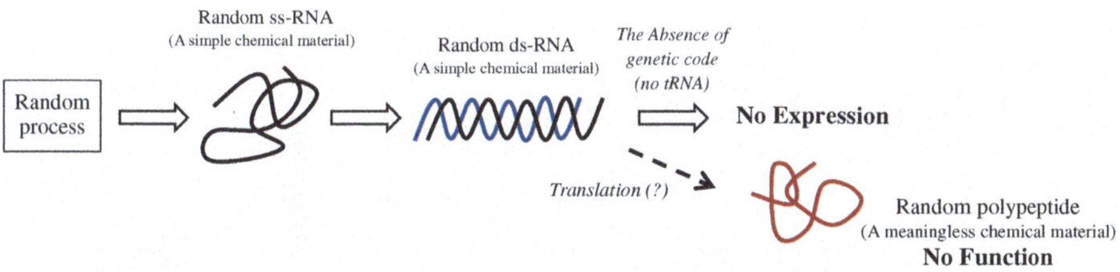

Fig. 3.5 Single-stranded RNA (ss-RNA) produced by random joining of nucleotides should be a simple chemical material without any genetic information for protein synthesis. Similarly, ds-RNA, which is formed by complementary strand synthesis of the ss-RNA, is also a simple chemical substance, because base sequences on both strands of the ds-RNA should be random. Of course, protein cannot be synthesized with the ds-RNA in the absence of tRNA and genetic code. Furthermore, a random polypeptide produced from the ds-RNA should not have naturally any catalytic function, even if the polypeptide could be produced from the ds-RNA

3.2 Protein Structure Hypothesis

Thus, alternative hypothesis was provided by Dill to avoid the weakness of the amino acid sequence hypothesis (Dill 1990). That is the protein structure hypothesis described below. The protein structure hypothesis is an idea giving weight to tertiary structure of a protein, rather than to primary structure or an amino acid sequence itself, for formation of a functional protein. Therefore, in the hypothesis, it is assumed that all proteins with a similar and homologous tertiary structure should exhibit the similar function. According to the hypothesis, it can be considered that the main chains of all the proteins are folded into a similar water-soluble globular structure, if a ratio of hydrophobic amino acids to hydrophilic amino acids and the number of hydrophobic interactions in a protein are the same (Fig. 3.6). Based on the assumption, a calculation was carried out with ribonuclease as an example using a lattice model of a protein and the number of amino acid sequences, which are folded into a similar conformation, was estimated as 10^{120} (Dill 1990). Therefore, it is supposed that an active enzyme accidentally uses one protein structure in the structure space, which is composed of proteins with a homologous three-dimensional structure viewed from the lattice model.

In Quanta Magazine (Cepelewicz 2017), the recent investigation of Dill's group based on the lattice model has been introduced as follows.

> *Dill indicated in his recent paper how early biopolymers could have grown long enough to fold into useful shapes, which may change that. If it holds up, the model could re-establish the reputation of proteins as the original self-replicating biomolecule (*Dill 1990*). Dill developed it in 1985 to help tackle the "protein-folding problem," which concerns how the sequence of amino acids in a protein dictates its folded structure. His hydrophobic-polar (HP) protein-folding model treats the 20 amino acids as just two types of subunit, which he likened to different colored beads on a necklace: blue, water-loving beads (polar monomers) and red, water-hating ones (nonpolar monomers) (*Fig. 3.6*). The model can fold a chain of these beads in sequential order along the intersections of a two-dimensional lat-*

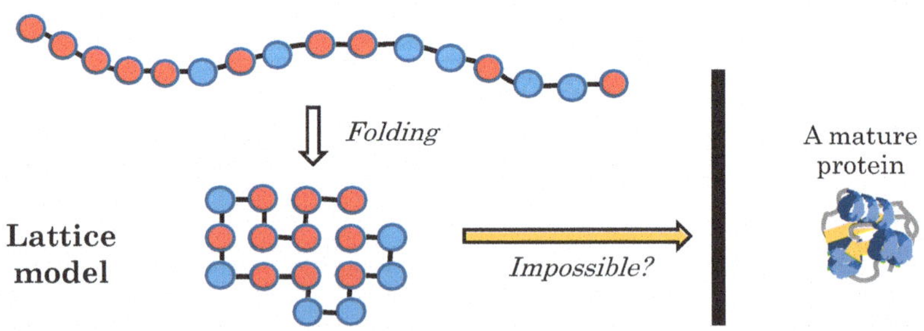

Fig. 3.6 Formation of protein structure according to the lattice model, which is considered by Dill's group (Dill 1990; Guseva et al. 2017). However, formation process of a real mature protein could not be explained by the lattice model, because both folding process of secondary structures or α-helix and β-sheet and the existence of peptide bonds in a protein are ignored in the model. Cartoon of the lattice model in Fig. 3.6 is drawn as tracing the figure appeared in the paper of Guseva et al. (2017)

tice, much like placing them on contiguous squares of a checkerboard. Which square a given bead ends up occupying depends on the tendency for the red, hydrophobic beads to clump together so that they can better avoid water.

Strengths of Protein Structure Hypothesis Certainly, Dill has indicated an important point of folding of polypeptide chain with two types of amino acids, hydrophobic and hydrophilic amino acids, as the Dill's model well explains the folding of a polypeptide chain into a water-soluble globular structure from a viewpoint of a basic protein science (Dill 1990; Guseva et al. 2017). Even the simple model treating only two types of hydrophobic and hydrophilic amino acids, could contribute to understanding of protein folding problem through hydrophobic interactions among hydrophobic amino acid residues of a polypeptide chain, until when alternative more realistic model is presented.

Weaknesses of Protein Structure Hypothesis However, there are many weaknesses in the protein structure hypothesis, because it would be difficult to explain rationally the formation process of real proteins, as described below. (1) Peptide bond (-CO–NH-) makes a plane to restrict free rotation around the C–N bond. The planarity of peptide bond largely contributes to the formation of secondary structures. However, the planarity of peptide bonds existing in actual proteins is completely ignored in the lattice model. (2) There are many intermediately hydrophobic and/or hydrophilic amino acids other than strongly hydrophobic amino acids and hydrophilic amino acids. (3) In addition, physico-chemical properties of amino acids are also variable as α-helix, β-sheet and turn/coil forming amino acids. Therefore, various properties of real proteins are not reflected in the lattice model, although, of course, basic property for folding of polypeptide chains is well explained in the model. (4) Furthermore, it would be also one of the weaknesses of the lattice model that amino acid sequence of a self-replicated protein, which is assumed in the hypothesis, could not be transferred to base sequence of a gene, as indicated by Crick (1958, 1970) (Chap. 2: Sect. 1.1). It means that formation process of mature and real proteins could not be explained with the Dill's lattice model (Dill 1990).

The reason, why the lattice model cannot be applied to creation of a real protein, is because the hypothesis oversimplifies the folding problem of a polypeptide chain, as that only hydrophobic (hydrophilic) interactions are considered as the dominant folding force. In addition, as well known, more than about 25% of amino acid residues are usually conserved among amino acid sequences of homologous proteins in a protein family with the same or a similar catalytic activity. The fact indicates that the homologous proteins have been produced from one common ancestor protein, and were not selected out independently from the sequence space or the structure space with the extraordinary large diversity. Therefore, the existence of conserved regions among homologous proteins is clearly inconsistent with the structure hypothesis. From such considerations, I would like to insist that the structure hypothesis cannot explain the origin of protein-II.

3.3 Short Peptide (Motif) Theory

In an introductory article on the short peptide theory of Dayhoff (Romero et al. 2016), it is described that polypeptides of variable lengths, essentially from 7 to 47 amino acids, can become functional through self-assembly and/or tandem fusion, and this may eventually yield globular, functional proteins. And thus, 50 years after Dayhoff's seminal paper, her hypothesis has been reinforced and validated, and now represents a general and proven model for the emergence of large, globular, and functional proteins from relatively short, simple peptides.

Then, I discuss the short peptide hypothesis, which is advocated based on the following features of protein.

As described in the paper introducing Dayhoff' theory (Romero et al. 2016), the most rudimentary functions make use of metal ions and/or ribonucleoside cofactors or cosubstrates such as ATP or NAD/H that probably emerged within the RNA world (Antonkine et al. 2009; Gibney et al. 2000). Accordingly, the most ancient peptide motifs mediate the binding of such cofactors and are omnipresent in the contemporary protein world.

Strengths of Short Peptide (Motif) Theory It looks like certain that modern proteins are assemblies of short peptides with characteristic second structures, as combinations of α-helix(s) and β-sheet(s), which are connected with turn/coil structures. The idea was obtained based on extensive analyses of a large number of three-dimensional structures of contemporary proteins. Therefore, the idea, in which the short peptide theory is advocated, seems also natural, because a large number of similar motifs can be found in databases of three-dimensional protein structures and a number of homologous proteins with the same catalytic activity or with a similar tertiary structure are found out in presently existing proteins.

Weaknesses of Short Peptide (Motif) Theory However, it is quite difficult and probably impossible to reconnect the short peptides with segments without conformation change upon the assembly and to restore the original catalytic center by connecting short peptides with turn/coil segments, as the short peptide theory expects (Fig. 3.7a). This would be obvious, taking it into consideration that an allosteric enzyme looses the catalytic activity even by a subtle conformation change upon binding of a ligand. Furthermore, the probability would be low that the substructures, which were cut off from a plural number of the same mature proteins, are rejoined by original combinations of the substructures (Fig. 3.7b). This also makes it difficult to restore the original structure of the protein. These indicate that mature proteins are not formed as expected by the short peptide hypothesis, but always constructed through the evolutionary process from an immature protein to a mature protein, as I expect (Sect. 4.8). In fact, Dayhoff, herself, stated in her paper that, however, whether these motifs display biochemical function as they are is yet to be examined and reconstructing simple proteins that comprise tandem repeats of these motifs is also an unmet challenge (Romero et al. 2016). The greatest weakness of

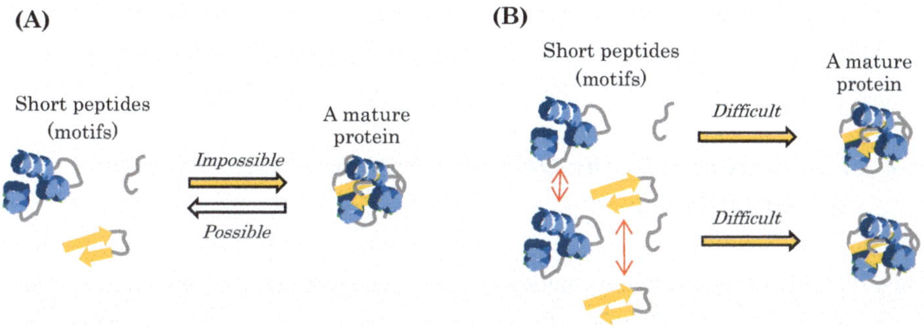

Fig. 3.7 (**a**) It is frequently considered that a mature protein is an assembly of short peptides or motifs and a mature protein could be divided into some substructures. However, it would be actually difficult to restore the objects to the original protein by assembling the motifs, because all the motifs could not be reconnected with turn/coil structures at the right positions without conformation changes. (**b**) Furthermore, it would be also quite difficult to rejoin the substructures by combinations of the original protein, which were cut off from a plural number of the same mature proteins. This implies that it would be impossible to restore the original mature protein. In addition, it is much more difficult for short peptides or motifs to retain original structures in a complex formed by joining of substructures among some different proteins

the theory would be that the protein structure formation started from analysis of tertiary structures of modern mature proteins (Chap. 2: Sect. 1.3.3).

Furthermore, some weaknesses of the short peptide theory are enumerated below.

(1) Synthesis of short peptides: It is not explained in the short peptide hypothesis on the origin of protein-II, how short peptides with an amino acid sequence could be produced through random process.
(2) Transfer of amino acid sequence to genetic sequence: It would be also one of great weaknesses of the short peptide hypothesis that amino acid sequences of short peptides could not be transferred to the respective base sequences necessary to synthesize the whole protein structure (Crick 1970).
(3) Low diversity of motif assembly: The following matters must be pointed out as one of weaknesses of the short peptide hypothesis. When protein structures are classified by a partial structure or a motif as assembly of secondary structures, α-helix(s), β-sheet(s) and turn/coil(s), which are observed in modern proteins, a large number of similar motifs should be observed. That is, diversity of partial structures of protein is quite lower than amino acid sequence diversity, because the number of secondary structures in one protein composed of 100 amino acids should be about 10, if it is assumed that the respective secondary structures are composed of 10 amino acids. The diversity of partial structures of protein is simply calculated as $3^{10} = {\sim}10^5$, since the number of types of the secondary structures are only three. Furthermore, the diversity becomes much lower as $2^{(10-5)} = {\sim}32$, assuming that the motifs are composed of two secondary structures, α-helix and β-sheet, except turn/coil structures connecting the two motifs and one turn/coil structure locating at one end. Therefore, it seems to me that

there is a large defect in the short peptide hypothesis, in which the accidental coincidences among partial structures or motifs of proteins are overvalued.

3.4 The Reason Why the Origin of Protein-II Has Not Been Solved Still Now

Only one side of the protein formation could be well explained by the three hypotheses, the amino acid sequence hypothesis, the protein structure hypothesis and the short peptide or motif hypothesis, which were proposed with the aim to solve the problem on the origin of protein-II. Inversely stating that, any idea has not succeeded to explain a reasonable formation process of a mature protein. The main reason, why the origin of protein-II cannot be explained by the three hypotheses, would be because many other researchers have considered the creation process of an EntNew protein without noticing the significance of the protein 0th-order structure. Other reasons, why the formation process of an EntNew protein cannot be explained by the three hypotheses, are as follows.

(1) Random process, which necessarily proceeded in formation process of an EntNew protein, is not considered in the respective three hypotheses.
(2) In the three hypotheses, the origin of protein-II has been explored as affected too strongly by the features of extant proteins, as amino acid sequence and three-dimensional structure.

Therefore, other researchers have not noticed that an EntNew protein was and has been actually formed through a quite different route from those expected by the three hypotheses. That is the protein 0th-order structure or a special amino acid composition as [GADV]-amino acids, which is the key concept of this book or the GADV hypothesis, which I have proposed (Ikehara 2002, 2005, 2014). Only the protein 0th-order structure could open the door to the most important step to the creation of EntNew protein (Sect. 4.8). It can be understood by introduction of the protein 0th-order structure for the first time that EntNew mature proteins could be created from the respective immature proteins through gradual evolution driven by a number of amino acid replacements one by one, after the first ds-$(GNC)_n$ RNA was formed. The origin of protein-II and its evolutionary process based on the protein 0th-order structure is explained in detail in Sect. 4.8.

4 My Idea About the Origin of Protein

4.A The Origin of Protein-I: How Was the First Mature Protein Created?
If any idea explaining the origin of protein-I has not been proposed by other researchers, it becomes a problem how a functional water-soluble globular or a

mature protein could be produced through random process before the first gene was created or in the absence of any genetic function. Thus, the quite difficult problem, how the first mature protein was generated through a random process on the primitive Earth either directly or indirectly, remains unsolved. I have considered that the protein 0th-order structure or a specific amino acid composition, in which an immature but water-soluble globular protein with some flexibility can be produced even through random process, holds the key for solving the problem (Ikehara 2014). Inversely, I even consider that the difficult problem never be solved, if the protein 0th-order structure is taken into consideration. I describe what protein 0th-order structure is and why the difficult problem can be solved under the protein 0th-order structure in detail in Sect. 4.2. The first necessary condition for generating a mature protein is to produce an immature protein with some flexibility, because the mature protein could be created from the immature protein through evolutionary process.

4.1 Chemical Evolution

Before the protein 0th-order structure is explained in Sect. 4.2, I would like to discuss first what kinds of organic compounds could be produced and accumulated during chemical evolution, in order to confirm whether [GADV]-amino acids and nucleotides, which are required to synthesize [GADV]-protein and RNA, respectively, could be accumulated with prebiotic means at large amounts on the primitive Earth.

Stanley L. Miller and Leslie E. Orgel described in their book (1974) as follows.

> *Amino acids were the first biologically interesting organic compounds to be identified as products formed under simulated primitive earth conditions. This was partly because amino acids were more easily identified at the time than purines and pyrimidines, and partly because they were formed in a single continuous set of operations under reasonably plausible prebiotic conditions.*

The description about synthesis of amino acids under the simulated primitive Earth conditions is quite plausible. On the contrary, it would be actually impossible to synthesize a sufficient amount of nucleotides with prebiotic means as so called "Miller's experiments", because the synthetic amount of nucleotide (AMP) can be estimated as roughly 10^{-6} of Gly (2 C atoms), taking the number of carbon atoms of nucleotides (AMP = 10 C atoms) into consideration (Fig. 3.8). Therefore, everyone could agree that the amounts of nucleotides accumulated were quite small, even if nucleotides were produced through prebiotic means or under similar conditions carried out by Miller (Miller and Orgel 1974). Furthermore, it is easily speculated that it would be also quite difficult to synthesize nucleotides at large amounts with catalytic reactions on surfaces of minerals as pyrite and clay as well as hydrothermal vents. These imply that nucleotides, which were required to synthesize RNA, should be produced through other routes than the usual prebiotic means.

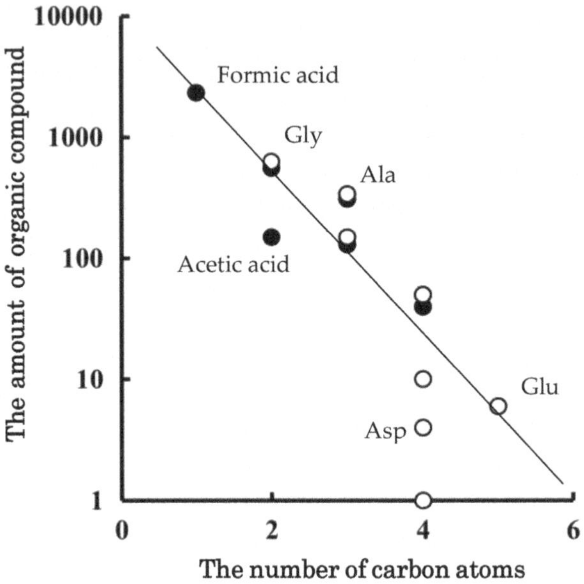

Fig. 3.8 Synthesis of organic compounds with the electric discharge experiments. The amount of an organic compound synthesized was plotted against the number of carbon atoms of the compound. The data used for the plots were borrowed from Table 8.1 published in the book of Miller and Orgel (1974). Amino acids and organic acids are shown with white and black circles, respectively.

4.1.1 The First Life Should Emerge with Organic Compounds Accumulated at a Large Amount on the Primitive Earth

In the abstract of a recent review article of Biscans (2018), it is described as follows. "By approaching this enigma (of origin of life) from a chemistry perspective, the goal is to define what series of chemical reactions could lead to the synthesis of nucleotides, amino acids, lipids, and other cellular components from simple feedstocks under prebiotically plausible conditions. Similarly, Patel et al. (2015) have described in their paper, "We show that precursors of ribonucleotides, amino acids and lipids can all be derived by reductive homologation of hydrogen cyanide and some of its derivatives and thus that all the cellular subsystems could have arisen simultaneously through common chemistry." Many researchers, especially chemists, generally consider that all organic compounds, which are necessary for the first life to emerge, must be synthesized with prebiotic means. Did the first life emerge as expected from the standpoint of chemical synthesis of organic compounds? However, quite small amounts of organic compounds with a large number of carbon atoms, especially as nucleotides, should be exhausted at one time after all before the emergence of life, even if nucleotides could be synthesized with prebiotic means.

Furthermore, one paper was published by Furukawa et al. (2019), describing that ribose and other sugars were detected from primitive meteorites including Murchison meteorite. However, ribose is still an intermediate for nucleotide synthesis. In addition, nucleotides could not be detected from the meteorites in the previous studies carried out by other researchers, suggesting that the amount of ribose (6.7 ~ 180 ppb

(Furukawa et al. 2019)) detected from the meteorites is too low to synthesize sufficient amounts of nucleotides and RNA, which are required to lead to the emergence of life. On the contrary, Gly, which is directly used to synthesize [GADV]-proteins, has been detected at a larger quantity (~3 ppm) than ribose from the meteorite (Koga and Naraoka 2017). Furthermore, the detection of ribose does not always support the RNA world hypothesis, because the following matters could not be explained reasonably. (1) How was a sufficient amount of RNA synthesized with the small quantity of ribose? (2) How was the genetic information for protein synthesis written into the RNA, even if RNA could be synthesized with the small quantity of ribose? (3) How was a metabolic system driven by RNA catalysts established in the RNA world? (4) How could the metabolic reactions driven by the RNAs be transferred to a proteinaceous metabolic system? I believe that all the four matters could not be explained from the standpoint of the RNA world hypothesis and, accordingly, the initial metabolism, which was driven by proteinaceous catalysts, should be established not after but long before the emergence of life.

4.1.2 The Problem of Homochirality on the Origin of Life

There is another serious problem in the origin-life research. That is the problem on why only L-amino acids and D-sugars as D-ribose have been used in extant organisms or how the homochirality of biomolecules could be attained on the primitive Earth. The problem also remains unsolved still now, although the homochirality is indispensable to form regular structures of protein or α-helix and β-sheet and of right-handed ds-DNA and RNA. However, I consider that the homochirality may be achieved during comparatively simple events on this planet, as described below.

1. Large amounts of simple organic compounds with a small number of carbon atoms as [GADV]-amino acids could be produced by prebiotic means as electric discharge into a reducing or even neutral primitive atmosphere (Fig. 3.8) (Miller and Orgel 1974; Cleaves et al. 2008). Thus, racemic [GADV]-amino acids first gradually accumulated on the primitive Earth for many years.
2. Successively, ([G] + L-[ADV])-amino acids might be preferentially but accidentally crystallized during concentration of the amino acids in depressions of rocks on seashores under sunshine on the primitive Earth. L-form excess [(G)ADV]-amino acids might remain in a depression after a large part of water was blown away from the depression by winds.

Some experimental results were previously reported, showing that differential crystallization could accidentally occur between racemic amino acid mixtures during concentration of the amino acid in aqueous solution (Kojo 2010; Breslow and Levine 2006). Kojo has published the results showing that L-form amino acids in racemic amino acids could be co-crystallized with L-Asn (Kojo 2010). Therefore, it can be supposed that homochiral [(G)ADV]-amino acids could be obtained by co-crystallization with L-Asp ammonium salt, which could exhibit a similar effect to

L-Asn. Such important experimental results for solving the homochirality problem of biomolecules have been curiously unnoticed until now.

3. Thereafter, homochiral immature [(G)ADV]-proteins as aggregates of [GADV]-peptides could be formed by repeated wet-drying cycles.
4. D-sugars as D-ribose could be synthesized preferentially by catalytic reactions on active sites of the immature but homochiral L-form [(G)ADV]-proteins, because a high reactivity for synthesis of D-sugars could be expressed on the [GADV]-proteins composed of homochiral L-form [(G)ADV]-amino acids. Formation of homochiral L-form [(G)ADV]-proteins determined the preferential usage of D-sugars as D-ribose.
5. Thus, the appearance of homochiral L-form [(G)ADV]-amino acids upon differential crystallization between racemic amino acids determined all homochiralities of biomolecules, although, of course, L-form excess amino acids, which were introduced from space (Pizzarello and Weber 2004; Pizzarello 2006), and the circularly polarized light in space (Bonner 1991) also might contribute to the formation of homochiral biomolecules. [GADV]-amino acids and [GADV]-proteins described in this book usually mean homochiral L-form amino acids and proteins, if especially not specified.

4.2 [GADV]-Amino Acids Could Be Produced at Large Amounts Through Prebiotic Means

The bottom-up approaches in the origin-life research, which are usually carried out by experiments, are undoubtedly important to understand how life emerged on the primitive Earth. Indeed, the important insight that [GADV]-amino acids could accumulate at large amounts on the primitive Earth was obtained through experiments.

As shown in so called "Miller's experiments" (Miller and Orgel 1974) and other similar experiments with prebiotic means (Cleaves et al. 2008), it has been confirmed by experiments that organic compounds as [GADV]-amino acids can be synthesized under random processes from simpler inorganic compounds, as CO_2, H_2O and N_2, and, therefore, that [GADV]-amino acids would accumulate at large amounts on the primitive Earth.

In facts, Higgs (2009) and van der Gulik et al. (2009) have reported in their papers that [GADV]-amino acids could accumulate on the primitive Earth at large amounts with prebiotic means, such as Miller's type electric discharge experiments, hydrothermal vents and chemical analysis of meteorites. Therefore, the conclusion that [GADV]-amino acids accumulated at large amounts on the primitive Earth is supported from the standpoint of chemical evolution. Next problem is whether [GADV]-protein could be produced through random process in the absence of any genetic system. Then, I explain my idea on the origin of protein-I in the next section.

Table 3.1 Three conditions for producing proteins before creation of the first gene or in the absence of gene

1. The first protein must be produced in the absence of gene, because nucleotides and RNA could not be produced at large amounts with prebiotic means.
2. The first protein must be produced through random process, because the proteins cannot be designed in advance.
3. The first protein must be not a mature protein but an immature water-soluble globular protein produced under protein 0th-order structure, because any mature protein never be produced in the absence of gene.

4.3 Immature [GADV]-Protein Could Be Produced Through a Random Process

As described in Chap. 2: Sect.1.2, it is important to understand for considering about the origin of protein, how the first mature protein was created through random processes or without gene on the primitive Earth, because both structure and function of the first protein cannot be designed in advance (Table 3.1). On the contrary, I have insisted thus far that the origins of gene and protein can be reasonably explained, if the evolutionary pathways from an immature protein to a mature protein are considered based on the protein 0th-order structure (Sect. 4.6) (Ikehara 2014). One of the protein 0th-orders structure or [GADV]-amino acids, which satisfy the four conditions for formation of water-soluble globular protein, made it possible to produce an immature but water-soluble globular protein in the absence of gene (Fig. 3.9) (Ikehara 2009, 2014). That is, [GADV]-proteins with a weak catalytic activity could be produced by random joining of [GADV]-amino acids under the protein 0th-order structure at a high probability, if the protein is an immature water-soluble globular protein with some flexibility (Table 3.1).

4.4 [GADV]-Amino Acid Composition Is One of Protein 0th-Order Structures

Here, I explain a protein 0th-order structure or a specific amino acid composition, in which an immature protein can be produced by random joining of amino acids at a high probability (Fig. 3.9). One of the protein 0th-order structures is the amino acid composition containing [GADV]-amino acids at roughly equal amounts. The reason, why the amino acid composition is one of protein 0th-order structures, is as follows (Ikehara 2002; Ikehara et al. 2002).

The average indexes of four fundamental properties of water-soluble globular proteins, (hydrophobicity/hydrophilicity (hydropathy), three secondary structure formabilities (α-helix, β-sheet, turn/coil)), were obtained by calculation with amino acid composition of a protein and the respective amino acid indexes given in Stryer's

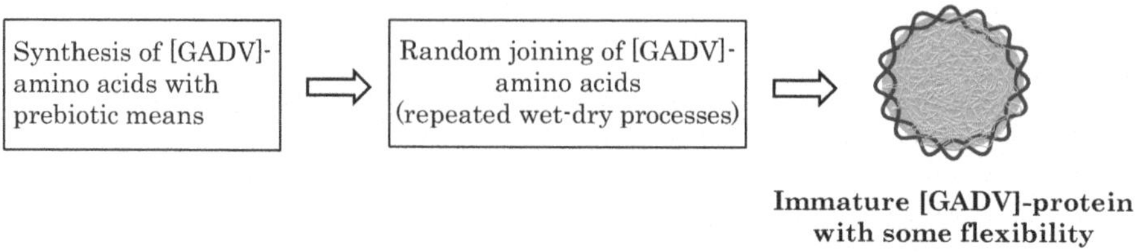

Fig. 3.9 Formation of an immature protein with some flexibility in the absence of genetic function. The reason, why peptides synthesized by random joining of [GADV]-amino acids could aggregate to form immature but water-soluble globular protein, is because the protein containing [GADV]-amino acids at roughly equal amounts can satisfy the four minimum conditions (hydropathy, α-helix, β-sheet and turn/coil formabilities), which are necessary to form water-soluble globular structure

text book (Berg et al. 2002), according to the following equation (Ikehara 2002; Ikehara et al. 2002). $I(x)_t = \Sigma_{a=1\sim 20} I(x)_a n_a/n_t$, where $I(x)_t$, $I(x)_a$, n_a and n_t are the total index of a protein, the index for each amino acid, the number of the respective amino acids and the total amino acid number in the protein, respectively. The four indexes were obtained by using water-soluble globular proteins encoded by seven microbial genomes with different GC content (*Mycobacterium tuberculosis* (GC = 65.6%), *Aeropyrum pernix* (56.3%), *Escherichia coli* (50.8%), *Bacillus subtilis* (43.5%), *Haemophilus influenzae* (38.1%), *Methanococcus genitarium* (31.3%) and *Borrelia burgdorferi* (28.2%)) (Ikehara et al. 2002; Ikehara 2002).

It was found that the four protein indexes were roughly constant, even if amino acid composition largely varied according to the change of GC content of a gene (Ikehara 2002; Ikehara et al. 2002). The four average indexes were as follows; hydropathy (−1.51 ± 0.38), α-helix (1.03 ± 0.03), β-sheet (1.00 ± 0.02), turn/coil (0.96 ± 0.05) formabilities (Ikehara et al. 2002). Therefore, it is expected that a polypeptide chain should be folded into a water-soluble globular structure, if all the four indexes of the imaginary protein fall within the respective ranges. It has been confirmed that proteins having the [GADV]-amino acid composition satisfies the four average protein structure indexes, when the four [GADV]-amino acids are contained roughly at one fourth each (Fig. 3.10). As described above, the protein indexes were calculated with amino acid composition, meaning that all proteins produced by random joining of amino acids in the same amino acid composition have the same protein structure indexes and, therefore, all [GADV]-polypeptide chains containing [GADV]-amino acids roughly at one fourth each should be folded into a similar water-soluble globular structure (Fig. 3.10), although the respective amino acid sequences are quite different from each other.

Therefore, the first meaningful chemical compounds, which could be produced through a random process, were only the immature water-soluble globular [GADV]-proteins. In other words, the process to the emergence of life could be initiated by formation of the immature [GADV]-proteins. Inversely stating this, life might not emerge on the primitive Earth, if the [GADV]-proteins or [GADV]-amino acids did not exist in this world. Another reason, why a meaningful protein or immature protein could be produced by random joining of amino acids, is because amino acids

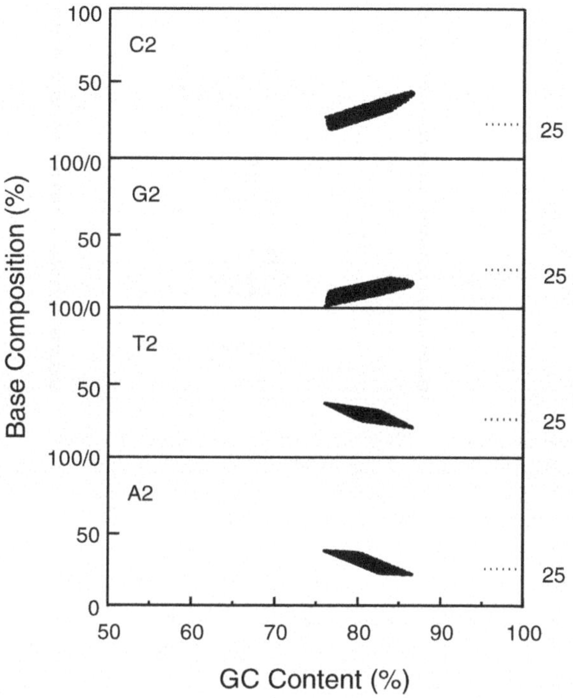

Fig. 3.10 Dot representation of computer-generated base compositions at the second base position in the codon. The base compositions were selected by determining whether or not an imaginary protein generated under the GNC coding system satisfies the four structural conditions (hydropathy, α-helix, β-sheet and turn/coil formabilities) for water-soluble globular structure formation

themselves are the respective functional units, unlike the case of gene that codons composed of triplet bases are functional units.

4.5 From Production of Immature Protein to Formation of Double-Stranded (GNC)$_n$ Gene

As described above, immature proteins could be formed by random joining of [GADV]-amino acids through repeated wet-drying processes without involvement from any other members, or cell structure, metabolism, tRNA, genetic code and gene, which constitute the fundamental life system. However, there is another difficult problem, which is necessary to solve the origin of protein, how the immature protein could evolve to a mature protein like a precision polymer machine, because any immature protein itself never directly evolve to the corresponding mature protein (Fig. 3.11). Then, only an outline of the evolution process is explained here, through which the immature protein could be matured, because the process of an immature protein to a mature protein is intimately related to the origins of cell

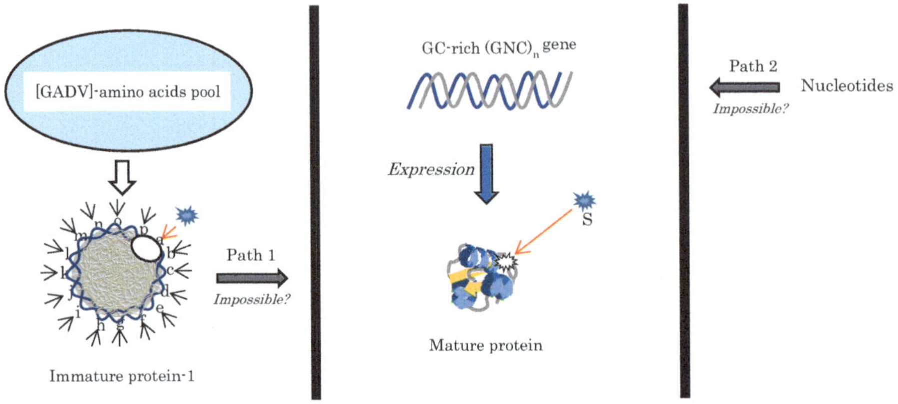

Fig. 3.11 An immature protein with some flexibility could be produced by random joining of [GADV]-amino acids. However, it is impossible to generate a mature protein directly from the immature protein (Path(way) 1). It is also impossible to create a gene encoding a mature [GADV]-protein by random joining of nucleotides (Path(way) 2)

structure, metabolism, tRNA and gene, and because the maturation process of an immature protein will be explained in detail in Chaps. 4, 5, 6 and 8.

4.5.1 Production of Immature Protein by Direct Random Joining of [GADV]-Amino Acids

(1) Immature [GADV]-protein-1 (actually, aggregates of [GADV]-peptides) was synthesized by random joining of [GADV]-amino acids, which accumulated at large amounts on the primitive Earth. The numeral "1" after immature [GADV]-protein means that the proteins produced are at a lower level, because the proteins are only aggregates of [GADV]-peptides synthesized by random joining of [GADV]-amino acids (Fig. 3.11). Peptide bond formation among [GADV]-amino acids would be preferentially progressed during repeated wet-dry cycles even in a circumstance on the primitive Earth containing various kinds of organic acids, amines etc., because the amino acids carry both positive (amino group) and negative (carboxyl group) charges in the molecules (Oba et al. 2005). However, non-natural amino acids such as D-amino acids, β- and γ-amino acids, 2-amino butylate and so on were also inevitably incorporated into the peptides, because it would be difficult to select out only ([G] + L-α-[ADV])-amino acids from the mixture.

4.5.2 Formation of Primitive Cell Structure or [GADV]-Microsphere and a Primitive Metabolic System

(1) A primitive cell structure or a [GADV]-microsphere was formed upon accumulation of immature [GADV]-proteins (see also Chap. 4) (Oba et al. 2005).

(2) A primitive metabolic system was also formed in the [GADV]-microcphere ([GADV]-protocell) or in [GADV]-protein world. In the [GADV]-protein world, [GADV]-amino acids, ribose, nucleotides and oligonucleotides could be synthesized by immature [GADV]-proteins (see Chap. 5: Sect. 3.1) (Ikehara under review).

4.5.3 Production of Immature Protein by Random Joining of Activated [GADV]-Amino Acids

After immature proteins-1 (Fig. 3.12) were produced by direct random joining of [GADV]-amino acids, immature proteins were produced by random joining of activated [GADV]-aminoacyl-AMPs ([GADV]-AMPs), which were synthesized with immature [GADV]-proteins and ATP. Successively, immature [GADV]-proteins were produced with activated [GADV]-aminoacyl-3'-ACC-5′ (Fig. 3.12).

4.5.4 Formation of Four AntiC-SL tRNAs Carrying [GADV]-Amino Acid and Synthesis of Immature [GADV]-Protein

(1) The first nonspecific anticodon stem-loop (AntiC-SL) tRNA was formed during repeated random synthesis-degradation cycles of oligonucleotides (Chap. 6: Fig. 6.9). After formation of four primitive nonspecific AntiC-SL tRNAs, immature

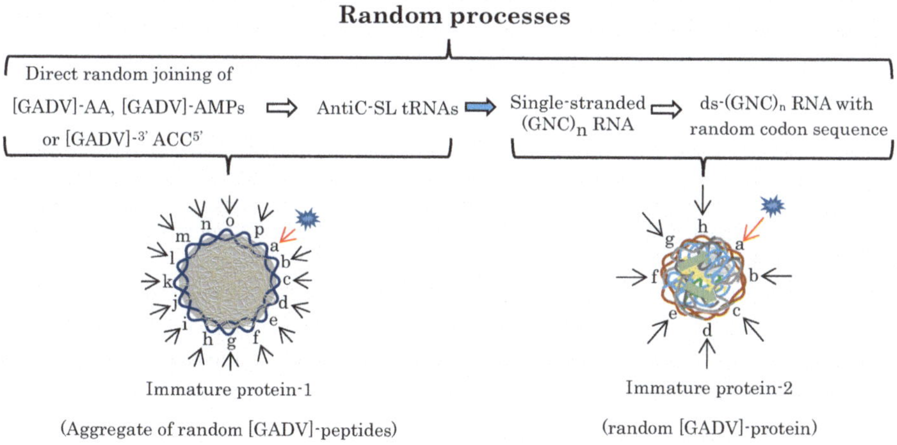

Fig. 3.12 Immature proteins-1 were formed by direct random joining of [GADV]-amino acids (AA), with [GADV]-AMPs or through primitive AntiC-SL tRNAs. Immature proteins-2 were produced from single-stranded (ss)-$(GNC)_n$ RNA or double-stranded (ds)-$(GNC)_n$ RNA. Both of the immature proteins-1 and -2 have pluripotency as shown by many short arrows around the protein. The cause of the pluripotency can be mainly attributed to the flexibility of the protein surface structure (shown by two wavy lines), on which the surface amino acids can be adjusted to a ligand

[GADV]-protein-1 was synthesized with the AntiC-SL tRNAs (Fig. 3.12) (Ikehara 2019). Use of the AntiC-SL tRNAs made it possible to synthesize immature [GADV]-proteins containing Gly/Ala and Asp/Val at ratios of roughly 1 for the first time.

4.5.5 Creations of Single-Stranded $(GNC)_n$ RNA and Double-Stranded $(GNC)_n$ RNA

(1) The first single-stranded (ss)-$(GNC)_n$ RNA was formed by random joining of GNC anticodons in anticodon loops of four nonspecific AntiC-SL tRNAs. Use of the ss-$(GNC)_n$ RNA enabled to produce many immature [GADV]-proteins-2 with substantially random but the same amino acid sequence because of the random GNC codon sequence on the ss-RNA (Fig. 3.12).

Here, the numeral "2" of immature [GADV]-proteins-2 means that the proteins produced are at a higher level than immature [GADV]-protein-1, because the proteins with a high molecular weight should be synthesized under random $(GNC)_n$ codon sequence. Differences of immature [GADV]-protein-1 from immature [GADV]-protein-2 is shown in Fig. 3.12 and fundamental properties of the three proteins, (1) mature protein, (2) immature protein-2 and (3) immature protein-1, are compared in Table 3.2, in order of proteins from at a higher level to a lower level.

(2) Successively, the first double-stranded (ds)-$(GNC)_n$ RNA could be produced by complementary strand synthesis of the first ss-$(GNC)_n$ RNA. After the acquisition of the first ds-$(GNC)_n$ RNA, an immature [GADV]-protein produced from either strand of the first ds-$(GNC)_n$ RNA could evolve to a mature [GADV]-protein for the first time (Fig. 3.13).

Table 3.2 Fundamental properties estimated of (**a**) a mature protein synthesized from codon sequence of a $(GNC)_n$ gene, (**b**) an immature protein-2, which was produced by expression of the first ss-$(GNC)_n$ RNA or on either one of two strands of the first random ds-$(GNC)_n$ RNA, and (**c**) an immature protein-1 produced by direct random joining of [GADV]-amino acids or activated [GADV]-amino acids

Properties	(a) Mature protein	(b) Immature protein-2	(c) Immature protein-1
Synthetic method	From $(GNC)_n$ gene	From $(GNC)_n$-NSF(a)	Random joining
Amino acid sequence	Ordered sequence	Essentially random	Random
Hydrophobic Core	Strong	Mediate	Weak
Compactness	Compact	Weakly swollen	Swollen
Rigidity	Rigid	Flexible	Quite flexible
No.of catalytic sites	1	Many (pluripotent)	Quite many (pluripotent)
Catalytic activity	High	Low	Quite low

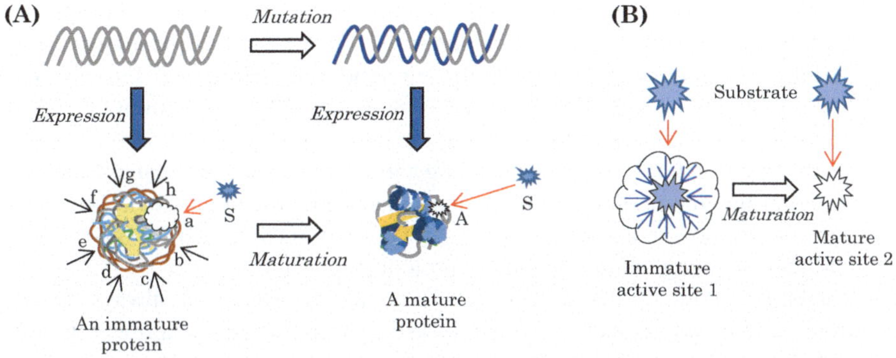

Fig. 3.13 (**a**) A mature protein is always created not directly but indirectly as a result of maturation (white bold arrows) from an immature protein with a low catalytic activity, which was produced through expression of one of $(GNC)_n$ sequences (gray wavy lines) of the first ds-$(GNC)_n$-RNA, to a mature protein produced from the mature gene (blue wavy line). Accompanied by the maturation of the immature protein, one sequence became a gene (blue wavy line) and the other became an antisense sequence (gray wavy line). (**b**) Formation of a mature active site on a protein. The site is always formed as a result of cumulative amino acid replacements, which were selected out from random amino acid replacements caused by random base substitutions of the genetic sequence, when the active site changed to fit the substrate more closely than before. Blue arrows surrounding the substrate show structural change of the immature active site 1 with a large flexibility induced upon binding with the substrate

4.6 *The Formation of the First ds-$(GNC)_n$ RNA Led to Creation of the First Mature [GADV]-Protein*

In order to elucidate the origin of protein-I, it is important to note that it is impossible to create a mature or a complete protein like a precision polymer machine at one stroke through random process on the primitive Earth (Fig. 3.11). However, the formation of the first ds-$(GNC)_n$ RNA made it possible to evolve gradually from an immature protein, which was synthesized through expression of a random codon sequence of ds-$(GNC)_n$ RNA, to a mature protein, as can be seen in Fig. 3.13.

4.6.1 Creation of a Mature [GADV]-Protein Was Triggered by the Formation of ds-$(GNC)_n$ RNA

(1) After formation of the first ds-RNA carrying random $(GNC)_n$ sequences, an immature protein, which was produced by expression of either one of two random $(GNC)_n$ sequences of the ds-RNA, could evolve to a mature protein as acquiring a higher activity step by step than before, if a weak but necessary activity was detected on the immature protein (Fig. 3.13).

(2) Thus, the first RNA gene encoding a mature protein was created through the evolutionary process from an immature [GADV]-protein-2 to a mature protein, as base substitutions are repeatedly but randomly introduced into the ds-RNA. In parallel, the ds-RNA evolved as memorizing base substitutions, which improved functionality of the evolving immature protein (Fig. 3.13b).
(3) As the first $(GNC)_n$ RNA gene encoding a mature protein was created from one strand of the first ds-RNA, GNC codon sequence on the other strand became a nonstop frame on antisense strand of GC-rich $(GNC)_n$ gene (GC-$(GNC)_n$ NSF(a)) encoding another immature protein (Fig. 3.13a).
(4) Thereafter, another EntNew gene encoding a mature protein was always created from a codon sequence on GC-$(GNC)_n$ NSF(a) encoding an immature protein-2.

Note that the creation process of a mature protein from an immature protein always progressed together with the creation process of an EntNew gene in parallel. Thus, the formation process of a mature protein cannot be discussed independently of evolutionary process of the corresponding gene (Fig. 3.13).

In conclusion, thereafter, even nowadays, formation process of an EntNew mature protein is carried out by an evolutionary process from an immature protein to a mature protein if necessary, as raising a catalytic activity upon amino acid replacements of the immature protein, which is produced by expression of a non-stop frame on antisense strand of modern GC-rich gene (GC-NSF(a)) (Ikehara et al. 1996). This is the only one way generating an EntNew mature protein.

4.7 Immature Protein Is Pluripotent

4.7.1 Evidence Showing That an Immature Protein Is Pluripotent

As described above, for creation of one EntNew mature protein, it is indispensable to detect a weak but sufficiently high activity on an immature protein (Fig. 3.14a). Therefore, all catalytic activities, which are necessary for the first life to emerge, must be contained in the whole group of immature proteins with some flexibility (Fig. 3.14b). This means that an extraordinary large number of catalytic activities must appear on one immature protein, because any activity never be detected on one protein as described in detail in the following sections, if one activity corresponds to one immature protein as like a mature protein. That is, one key for solving the origin of life is protein 0th-order structure, in which immature proteins could be produced by random joining of amino acids and the other key is pluripotency of an immature protein, which is enabled by a flexible surface structure of the immature protein and by adjustment of the surface structure to a large number of organic compounds. The fact, that both life emerged on the primitive Earth and diverse organisms inhabit on the present Earth by creating EntNew proteins when necessary, demonstrates that both immature protein-1 and immature protein-2 were and are pluripotent (Fig. 3.14b).

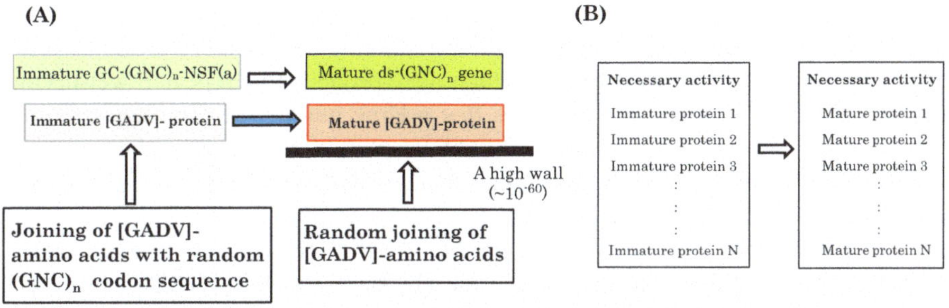

Fig. 3.14 (**a**) A mature [GADV]-protein cannot be produced directly by random joining of [GADV]-amino acids because of a high wall of amino acid sequence diversity of $4^{100} = {\sim}10^{-60}$. Therefore, both the first mature protein and mature proteins after the emergence of life had to be created through evolutionary process from an immature protein, which was synthesized by expression of one strand of the first ds-$(GNC)_n$-RNA or a GC-$(GNC)_n$-NSF(a) with random GNC codon sequence. (**b**) On the other hand, all metabolic reactions in the initial metabolism were catalyzed by the respectively corresponding mature proteins after the first ds-RNA formation. This means that all necessary catalytic activities had appeared on the surfaces of a whole group of immature proteins, because all mature protein exhibiting a necessary catalytic activity must be created from the corresponding immature protein. The situation could be realized, only when an immature protein is pluripotent

Therefore, it would be unquestionable that even the immature [GADV]-proteins, which were produced by random joining of [GADV]-amino acids, have a large number of catalytic activities. So, I would like to give the immature protein a name, immature pluripotent protein or iPP protein. However, the catalytic activity of the iPP protein should remain low, because the protein could not raise the activity to a high level as the protein itself did not possess a memorizing ability of amino acid replacements and, therefore, the protein could not evolve and the activities could not be matured in the absence of ds-RNA/DNA (Fig. 3.11).

(1) **Theoretical evidence**

1. The estimated molecular number of [GADV]-proteins in a test tube: It is easily understood that an immature protein is pluripotent, from relation between the molecular number of [GADV]-random copolymers ([GADV]-proteins) in a test tube and a sequence diversity of the [GADV]-random copolymers. Then, calculate the molecular number of 100 mer [GADV]-polypeptides, which were produced by repeated wet-drying cycles of 10 μmole each of [GADV]-amino acids in 1 ml, is estimated as roughly $\sim 2 \times 10^{17}$ [GADV]-polypeptide molecules. On the other hand, the sequence diversity of [GADV]-protein composed of 100 of [GADV]-amino acids is calculated as $4^{100} = {\sim}10^{60}$. If it is assumed that one immature [GADV]-protein having a unique amino acid sequence exhibits one specific catalytic activity like as a mature protein, a specific catalytic activity never appear in a test tube containing the immature [GADV]-proteins, because the sequence diversity ($\sim 10^{60}$) is much larger than the molecular number of immature [GADV]-proteins in a test tube ($\sim 2 \times 10^{17}$), and, therefore, the appearance probability is estimated as ($\sim 2 \times 10^{17} / \sim 10^{60} = \sim 2 \times 10^{-43}$).

Nevertheless, it was confirmed by the experiments showing that various catalytic activities were easily detected in a test tube containing the [GADV]-proteins, actually aggregates of [GADV]-peptides, which were produced by repeated wet-drying cycles (Oba et al. 2005). This indicates that a large number of catalytic activities should be expressed on one immature [GADV]-protein-1, because, otherwise, the experimental results cannot be explained reasonably, considering the appearance probability of a catalytic activity ($\sim 10^{-43}$).

2. Estimation of the number of possible catalytic centers on one immature [GADV]-protein: Next, consider the reason why one immature [GADV]-protein-1 can exhibit so large number of catalytic activities or is pluripotent.

1. Consider how many possible active sites appear on the surface of one immature [GADV]-protein from the number of combinations of surface amino acids. First, it is estimated that about 40 amino acid residues should be appear on the surface of one protein composed of 100 amino acids, based on the theory of spherical virus structure (Caspar and Klug 1962). Next, the number of possible active sites is estimated as at least ~1000 from possible combinations of three and four amino acids on the surface of one [GADV]-protein. Therefore, even one [GADV]-protein has a possibility exhibiting a large number of catalytic activities on the surface of the [GADV]-protein.
2. Furthermore, swinging of the 40 surface amino acid residues, which is attributed to the flexibility of an immature protein, increases the number of possible candidates for the active sites. The reason is because the number is estimated as roughly 10^{40}, if it is assumed that one amino acid residue could be occupied 10 different positions by wobbling of the surface.
3. It is also expected that the wobbling of the surface amino acids of an immature protein could made it possible to adjust the structure of a possible catalytic site to many organic compounds like as the "induced-fit" mechanism of modern mature enzyme (Fig. 3.13b) (Koshland 1995). Accompanied by the wobbling of the surface amino acids, an active site can receive many different organic compounds by wrapping up the organic compound as searching for the most stable state of the complex.
4. On the other hand, the large wobbling of immature protein should cause to lower the activity. However, the low activity of the immature protein was not a problem, because even the low activity would be enough in the absence of the same catalytic activity on the primitive Earth. The weak catalytic activity of such an immature protein triggered the creation of a mature protein after ds-$(GNC)_n$ RNA was formed (Fig. 3.13a). Successively, an active site with a higher catalytic activity was formed as piling up amino acid replacements to adjust the site to the corresponding substrate (Fig. 3.13b). Thus, a mature protein was always created from an immature iPP protein encoded by a nonstop frame on antisense strand of ds-GC-$(GNC)_n$ gene (ds-GC-$(GNC)_n$-NSF(a)).

In short, immature [GADV]-proteins-1 and -2 could bind with organic compounds and could catalyze them by adapting the surface structure of the protein to the organic compounds owing to the flexible structure of the protein, although, of course, an immature protein for catalyzing an organic compound had not been prepared in advance. It would be better to consider that some elements, as a large number of combinations and wobbling of the surface amino acids of an immature [GADV]-protein, enabled to generate a huge number of possible catalytic centers with a weak but sufficiently large affinity to a reactant. Generation of such immature proteins with various catalytic activities through random processes was significant for both the origins of protein and life.

(2) **Experimental evidence**

1. Fox group's experiments: In 1970s, Fox et al. had already recognized that the first life might emerge as protenoid microspheres, which were produced by repeated wet-drying processes, although they did not study with [GADV]-protenoids (Fox et al. 1974). They described in their paper as follows.

In this paper, we report mechanisms by which two kinds of microparticle from protenoid (ptd) use one component of their environment, ATP, to produce two kinds of polymer molecule, polyamino acids and polynucleotides. (ATP with acidic ptd microspheres).

In addition, Fox et al., had also confirmed that the polymerization reaction could be catalyzed by acidic protenoid-microsphere (Fox et al. 1974). Their results indicate that acidic [GADV]-protenoids also could polymerize amino acids and nucleotides. It should take notice that Fox et al. used amino acid mixture containing of Lys and Arg, which it should be difficult to synthesize with prebiotic means, in their experiments of protenoid microsphere (Fox and Dose 1977). It is supposed that the reason is probably because the microspheres constructed with Lys expressed high catalytic activities. Therefore, it is expected that acidic [GADV]-microspheres, which were produced by repeated wet-drying cycles, also could synthesize polypeptide chain.

2. Ikehara group's experiments: Immature [GADV]-proteins-1, which were produced through random joining of [GADV]-amino acids, should express various catalytic activities, if the consideration described in Sect. 4.7.1 is correct. In facts, the studies along the line of the consideration were previously carried out by Ikehara's group. Oba et al. (2005) reported that [GADV]-proteins produced by repeated wet-dry cycles exhibited several catalytic activities, as β-galactosidase activity toward 4-methylumbelliferyl-β-D-galactoside, peptidase activity toward glycine-*p*-nitroanilide and protease activity against several peptide bonds of bovine serum albumin. Furthermore, our group has confirmed that [GADV]-proteins have also tRNA hydrolytic activity. I was so surprised at the facts that not many but all the catalytic activities examined were detected with the immature [GADV]-proteins-1. Therefore, at first I hesitated to submit the important experimental results as a manuscript (Oba et al. 2005). The reason is because the appearance probability was so much higher than what I expected at that time. However, I have recognized now that those were quite important results indicating that every immature protein should certainly be pluripotent.

4.B *The origin of protein-II:* How has an entirely new mature protein been created?

4.8 Creation of an Entirely New Mature Protein Through the First Double-Stranded $(GNC)_n$ RNA

4.8.1 No Mature Protein Could Be Produced Through Random Process

It is unquestionable that many proteins, which belong to the respective protein families, are derived from the most ancestral protein in the protein family as expected by gene duplication theory (Ohno 1970). This clearly indicates that many EntNew proteins or the first proteins of many protein families have been created thus far ever since the first $(GNC)_n$ gene encoding a mature protein was created. Therefore, the key problem for solving the second origin of protein-II is to understand the process how an EntNew protein, which does not show any meaningful homology with any other previously existed proteins, have been created.

On the other hand, it has been generally considered thus far that an EntNew mature protein should be generated accompanied by creation of the corresponding ds-DNA/RNA gene (Fig. 3.3), because gene but not protein plays a central role in production of protein under the modern genetic system (Sect. 3.1). One of the reasons, why the genetic information is required as the premise for creation of an EntNew mature protein, would be because it has been considered that any EntNew mature protein never be created in the absence of the genetic information (Table 3.3: Principle 1.1), as suggested by the "gene/replicator-early" theory. This is obvious from the fact that the origin of life has been discussed still now according to the RNA world hypothesis based on the "gene/replicator-early" theory. However, the creation of an EntNew mature gene or the first family gene encoding an EntNew mature protein should be, of course, carried out through random process or with not literally but essentially random codon sequence, because any EntNew gene cannot be designed for an EntNew mature protein in advance.

However, such an EntNew gene could not be created by random joining of nucleotides (Table 3.3: Principle 1.2). The prejudice that a gene must lead the creation of an EntNew protein, might prevent to make clear the process how an EntNew protein has been created. In addition, it seems to me that the second origin of protein-II has not been seriously considered even by many researchers working in the research

Table 3.3 Principles for creation of a mature protein

Principle 1.1. Mature protein cannot be created through random process at one stroke, because of the extraordinary vast amino acid sequence diversity, $\sim 10^{130}$.
Principle 1.2. Similarly, gene encoding a mature protein cannot be created through random process at one stroke, because of the extraordinary vast base sequence diversity, $\sim 10^{180}$.
Principle 2. Immature but water-soluble globular protein could be produced by substantially random joining of amino acids in a protein 0th-order structure
Principle 3. Therefore, a mature protein must be created through evolution from an immature protein with essentially random amino acid sequence, which is encoded by pan-GC-NSF(a), with the memorizing ability of ds-RNA/DNA.

field of protein science and, therefore, the origins-II of both protein and gene never be solved, even if RNA could be produced with prebiotic means, because RNA, which was produced by random joining of mononucleotides, should be a simple chemical material without genetic information for protein synthesis (Fig. 3.5). Another reason is because RNA produced through the random process is composed of a base sequence but not codon sequence, which is required to encode an amino acid sequence of a protein.

This is similar to that a polypeptide chain produced by random joining of amino acids not in a protein 0th-order structure is also only a chemical material. In addition, RNA could not express any genetic function before establishment of a genetic code or before formation of primitive tRNAs realizing the genetic code (Fig. 3.5). Thus, the second origin of protein-II has not been seriously discussed thus far, because it might be unconsciously considered that it is probably impossible to synthesize a mature protein through random process.

4.8.2 Every Mature Protein Was Always Created Through Maturation Process from an Immature Protein

It is of course essential to understand the mechanism, how an EntNew mature protein could be created, for solving the problem on the origin of protein-II. As described in Sects. 4.2, 4.3, and 4.4, for the purpose, it is indispensable to recognize that any mature protein must be created through the maturation process from an immature protein, which was produced through random process, ever since after the formation of the first ds-$(GNC)_n$ gene (Sect. 4.6.1). The amino acid sequence, which satisfies the two conditions, (1) starting from an immature protein and (2) production through random process, is encoded by pan-GC-NSF(a). Here, the word, pan-GC-NSF(a), means the three fields for creation of EntNew proteins, or three codon sequences encoding substantially random amino acid sequence, that is, (1) GC-NSF(a), (2) GC-$(SNS)_n$-NSF(a) and (3) GC-$(GNC)_n$-NSF(a), under universal or standard genetic code, SNS primitive genetic code and GNC primeval genetic code, respectively.

The mechanisms for creation of the respective EntNew mature genes/proteins under the three fields will be described and discussed in detail in Chap. 8, Sect. 3.3, because the creation mechanism of an EntNew protein can be understood as the creation mechanism of an EntNew gene encoding the EntNew protein. Creation process of an EntNew protein could be well understood only by considering evolution process from an immature protein encoded by a pan-GC-NSF(a) to a mature protein as raising a catalytic activity upon amino acid replacements of the immature protein, which was produced by expression of the pan-GC-NSF(a). Therefore, EntNew proteins have been always created from the corresponding immature proteins in the every genetic code era or GNC code, SNS code and the universal genetic code (Chap. 8: Fig. 8.9).

4.8.3 Creation of Mature Protein Was Always Led by Not Gene But the Protein Itself

Next, I would like to explain that the main motive force for creating an EntNew protein is not in gene but in protein itself, although support by a ds-RNA/DNA or a pan-GC-NSF(a) of a gene is, of course, secondarily necessary to create the EntNew protein. In the following sections, it is discussed with the word "ds-DNA" instead of "ds-RNA/DNA", unless otherwise specified. The reason is because it can be similarly discussed in both cases of ds-DNA and ds-RNA and also because it is considered that ds-DNA has been used as a genetic material for a long while after the first DNA gene was created.

It is clear that change of an amino acid sequence of an immature protein is triggered by a base substitution on ds-DNA gene, and that the evolutionary process of the protein is memorized as the base substitutions. Accordingly, it seems that the base substitutions of ds-DNA determine the evolutionary process of the immature protein. However, it would be more appropriate to consider that not ds-DNA but protein itself makes it possible to create an EntNew protein, because ds-DNA only memorize the amino acid replacements, which are induced by random base substitutions onto the ds-DNA. That is, during evolution of the protein, what kinds of base replacements should be memorized in the ds-DNA, is determined by the functional changes of the protein. In other words, although base substitution of ds-DNA certainly causes the change of the protein function, it is rightly determined by improvement of the function of the newly mutated protein. DNA never play a leading role in formation of an EntNew protein. That is, the random base replacements of the ds-DNA cannot determine the evolutionary process of the immature protein and the ds-DNA itself cannot create any new amino acid sequence of the protein evolving from an immature protein to a mature protein. During the evolution, the DNA encoding the evolving protein only plays an auxiliary role in the protein evolution through the ability of ds-DNA memorizing amino acid replacements. If the evolutionary process of the protein was traced later, as a result, it only seems that DNA or gene played a leading role in the evolutionary process of the protein.

The next example would make it understandable more easily that protein holds the initiative in its protein maturation. Assume a situation that a ball falls down an uneven slope during winds randomly blow. A wind would trigger for a ball to fall down the slope. However, the path traced by the falling ball on the slope is determined by both size and weight of the ball and by shape of depressions on the slope but not by the winds, which blow randomly. Accordingly, the trajectory of the falling ball is determined not by the randomly blowing winds, which correspond to the random base substitutions of ds-DNA, but by the falling ball, which corresponds to a protein, and by depressions on the slope, which corresponds to the change of catalytic activity.

Furthermore, as a matter of course, maturity of gene/protein, or whether the evolution of a gene or of the function of a protein has arrived at the final maturation point or not, is determined by maturity of the protein but not of the gene. This also indicates that not gene but protein is an main actor in the genetic system until

Table 3.4 Which, protein and gene, played the leading role (L.R.) in [GADV]-proteins? [GADV]-proteins were synthesized without gene during step 1~6 before the first gene was created. The leading role was apparently transferred from protein to gene, after creation of the first ds-$(GNC)_n$ gene

[GADV]-protein synthesis	L.R.
1. Direct random joining of [GADV]-amino acids	Protein
2. Random joining of activated [GADV]-aa-AMP	Protein
3. Random joining with [GADV]-amino acids bound with CCA	Protein
4. Random joining of [GADV]-amino acids through AntiC-SL tRNAs	Protein
5. [GADV]-protein synthesis according to ss-RNA with random $(GNC)_n$	Protein
6. [GADV]-protein synthesis according to ds-RNA with random $(GNC)_n$	Protein
7. [GADV]-protein synthesis with ds-RNA encoding a mature [GADV]-protein	Gene*

*mark indicates the apparent transfer from protein to gene

function of the protein reaches the optimal stage. All organisms living on the present Earth have used mature proteins, which have evolved to the respective optimal stages. After the protein reached the optimal stage, it simply seems that a gene determines the function of a protein or gene plays a leading role in producing protein. Thus, the leading role in the genetic system has been reversed formally from protein to gene (Table 3.4). We still tend to consider the creation processes of gene/protein with the gene-centered idea even before the leading role was apparently replaced from protein to gene, because we pay close attention to gene in the relation between gene and protein. The prejudice would have made it difficult to solve the origins of protein and life. Any gene (genetic information) encoding an EntNew mature protein never be created independently of protein because the fundamental life system has been established to produce proteins.

4.8.4 Formation of an Active Site of an Induced-Fit Type Mature Protein

Characteristics of the induced-fit model proposed by Koshland are enumerated to introduce his model in the paper (Koshland 1995) as follows.

1. *Koshland had advocated a theory to account for the specificity of enzymes.*
2. *He postulated that the essential functional groups on the active site of the free enzyme are not in their optimal positions for promoting catalysis.*
3. *When the substrate molecule is bound by the enzyme, the catalytic groups move to their favorable geometrical positions to form the transition state.*
4. *The enzyme molecule is unstable in this active conformation and tends to revert to its free form in the absence of substrate.*
5. *In the induced fit model, the substrate induces a conformational change in the enzyme which aligns the amino acid residues or other groups for substrate binding, catalysis or both.*

How was the amazing protein, which can bind with the substrate as proposed by the induced-fit model, created? In the induced-fit model, it is assumed that the

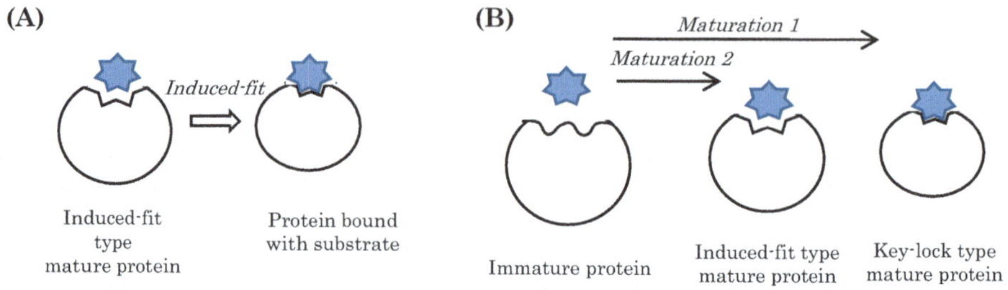

Fig. 3.15 (**a**) The active site of a free induced-fit type enzyme stands by a little wider than the position closely binding with the substrate. The substrate induces a conformational change of the enzyme to form the complex between the substrate and the active site in an optimal state. (**b**) A mature protein is always formed from an immature protein with some flexibility. The flexible immature protein is gradually matured as accumulating appropriate amino acid replacements, if a weak but necessary catalytic activity can be found on the surface of the immature protein. In some cases (case 1), the key-lock type mature protein is formed as binding closely with the substrate from the immature protein (maturation 1). In other cases (case 2), the induced-fit type mature protein with a little wider active site than the substrate is created through maturation process from an immature protein by terminating the maturation process just before formation of the key-lock type mature protein is completed (maturation 2). In both cases, the mature proteins are optimized according to the respective necessities

essential functional groups on the active site of the free enzyme are arranged at a little bit wider locations than their optimal positions for promoting catalysis. Then, the substrate induces a conformational change in the enzyme to fit the substrate to the amino acid residues closely (Fig. 3.15a). The reason, how the induced-fit type mature protein could be created, is well understood by considering an evolutionary process from an immature protein to the corresponding mature protein as described below (Fig. 3.15b).

The reason, how such an amazing mature enzyme could be created, is because, in the case of the key-lock type enzyme, the enzyme is created from an immature protein with a catalytic site, which is enough larger than the substrate, to the mature enzyme with a catalytic site, which fits closely to the substrate (Fig. 3.15; maturation 1). On the other hand, in the case of the induced-fit type enzyme, the enzyme is created from an immature protein to the mature enzyme similarly to the case 1. However, in the case 2, the maturation process is terminated to form a little wider active site than the substrate at just before the key-lock type enzyme is completed (Fig. 3.15; maturation 2). These would be the most optimal strategies for creation of the two types of mature proteins. The other way than the formation of a mature protein from an immature protein must be virtually absent. Inversely stating this, it also indicates that every mature protein has been always created from an immature protein.

Strengths of Protein 0th-Order Structure Hypothesis

Whether or not something assumed is valid as the origin relies on whether or not it can be explained that something assumed could be produced through random process and could evolve to the modern one. According to the conditions, it can be concluded that an immature [GADV]-protein, which was produced by random joining of [GADV]-amino acids in a protein 0th-order structure, is valid as the origin of protein-I, because both formation of the immature [GADV]-protein in the absence of gene and the evolution process from the immature protein to a mature protein can be reasonably explained based on the protein 0th-order structure, which was obtained through database analyses of microbial genes and proteins, as follows.

1. *Before formation of the first double-stranded (GNC)$_n$ RNA*: The first protein was an immature but water-soluble globular [GADV]-protein, actually aggregate of [GADV]-peptides produced by random joining of [GADV]-amino acids in the protein 0th-order structure, which were produced with prebiotic means and accumulated on the primitive Earth.
   Furthermore, creation process of a mature protein, which was produced after creation of the first ds-$(GNC)_n$ RNA, can be also understood with the protein 0th-order structure. That is, representative examples are enumerated as follows.
2. *Formation of the first mature [GADV]-protein*: The creation process of the first mature and rigid proteins can be explained based on the protein 0th-order structure. Namely, the first mature [GADV]-protein could be created through evolutionary process from an immature and flexible [GADV]-protein, which was produced by expression of the first ds-$(GNC)_n$ RNA encoding one of random [GADV]-amino acid sequences. The creation of a mature protein has been led by protein itself through functional promotion as supported by memorizing ability of ds-RNA
3. *After formation of the first double-stranded (GNC)$_n$ gene*: Furthermore, EntNew mature proteins were similarly created even in later eras of GNC code, SNS code and the universal genetic code, under the essentially same mechanism. The mature proteins were formed through maturation process from an immature protein with an essentially random amino acid sequence, which was produced by expression of pan-GC-NSF(a) under protein 0th-order structure. The immature proteins with some flexibility enabled to generate various amazing mature proteins with rigid structure.

Weaknesses of Protein 0th-Order Structure Hypothesis

As described in the above strengths of protein 0th-order structure hypothesis, I am considering that I could explain the two origins of protein-I and -II with protein 0th-order structure hypothesis. (1) The first one is "How was the first protein created? (2) The second one is "How were entirely new proteins generated under the three genetic codes or GNC code, SNS code and the universal genetic code?" However, it does not always mean that the creation process of a mature protein deduced by protein 0th-order structure hypothesis is correct, on the grounds that the creation processes can be reasonably explained or that the creation process could not be reasonably explained by the other ideas.

Therefore, it is quite important to confirm with experiments whether the creation processes of mature proteins deduced by the protein 0th-order structure hypothesis is correct or not. Thus, the greatest weakness of the hypothesis is that the creation processes of mature proteins from immature proteins have not been confirmed by experiments.

However, as described in Chap. 2; Sect. 1.3.1, it is important to first narrow down the events, which might happen on the primitive Earth, into ones which would occur at a high probability as much as possible with theoretical approaches, especially in the case of the origin of something. Therefore, it is important to try to make clear the quite difficult problems on the origin of protein with experiments after it was previously confirmed theoretically whether or not the origin of protein can be explained by a new concept and there are possible steps to evolve to modern protein as described in this Chapter.

5 Discussion

The conditions for concluding that something assumed is suitable to the origin of protein, are described below.

(1) The original thing must be one produced through random process (Fig. 3.16a-1, b-1, c-1).
(2) The most primitive thing among ones, of which features are similar to the modern one, should be the original one (Fig. 3.16b).
(3) If the assumed thing is dissimilar to the modern one, the dissimilar thing must be transformed to the most primitive and similar one (Fig. 3.16a-2).
(4) The most primitive thing with similar features to modern one must be able to evolve to the modern one (Fig. 3.16b-3).

It is discussed and confirmed whether the hypotheses, which were proposed by other researchers, and my own idea, are valid for the origin of protein as referring the conditions shown in Fig. 3.17.

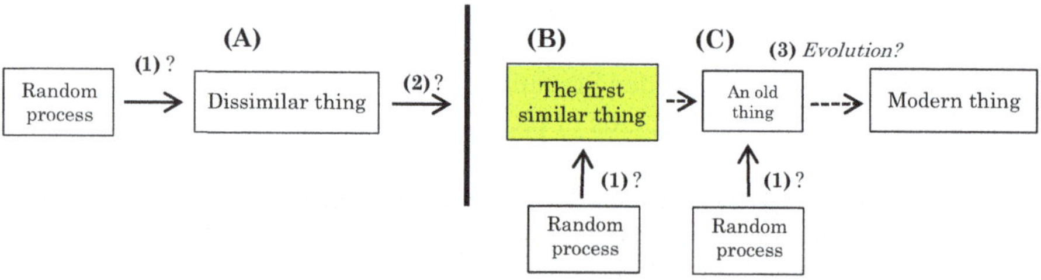

Fig. 3.16 The necessary conditions for being the origin of something. (**a**) When features of a primitive member assumed as the origin of something were dissimilar to the features of the modern one, it must be concluded that the assumed dissimilar thing is not the origin of the something, if the following three points could not be satisfied. (1) The primitive member with different features from modern one must be produced through random process. (2) The dissimilar thing must be transformed to the most primitive and similar thing to modern one, either directly or indirectly. (3) The most primitive thing transformed from the dissimilar thing must evolve to modern thing. (**b**) If the first primitive thing with features similar to modern one was assumed, the following three points must be satisfied. (1) The assumed thing must be produced through random process. (2) The assumed thing must be the most primitive one among things, which have similar features to modern one. And also (3) evolutionary pathway from the most primitive one to modern thing must be explained

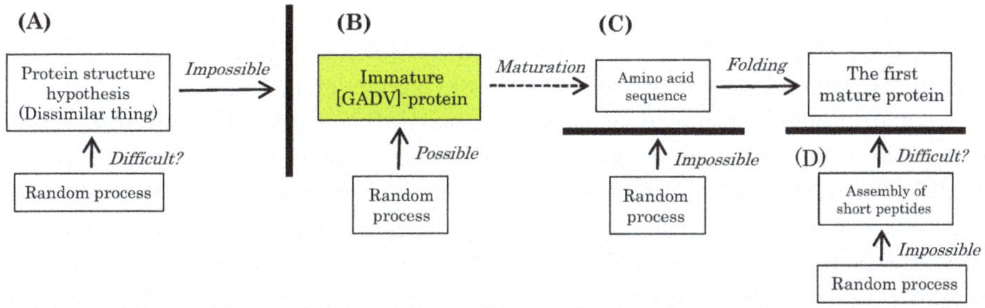

Fig. 3.17 It can be verified whether or not various hypotheses proposed by other researchers and by me are valid according to the necessary conditions for being the origin of something as shown in Fig. 3.16. (**a**) Protein assumed by protein structure hypothesis (Dill 1990; Guseva et al. 2017) is dissimilar to modern protein, because the protein, which is assumed based on the lattice model, is quite different from a real protein and also because it would be impossible to transform the assumed protein to a real protein. (**b**) Immature [GADV]-protein supposed by GADV hypothesis can be produced by random joining of [GADV]-amino acids and can evolve to modern protein (Fig. 3.18). (**c**) The amino acid sequence assumed by amino acid sequence hypothesis cannot be produced by random process (Crick 1958; Dill 1990). (**d**) It would be impossible to produce all short peptides, which are necessary to construct one mature protein, through random processes and also it would be difficult to construct a mature modern protein by connecting the short peptides with turn/coil peptides (Sect. 3.3)

Then, it can be judged whether or not the things assumed by other researchers are suitable to the origin of protein-II. (i) *Amino acid sequence hypothesis*: Polypeptide chain with an amino acid sequence, which can be folded into a mature water-soluble globular protein, could not be produced with prebiotic means on the primitive Earth. Therefore, the amino acid sequence hypothesis is incompatible with the condition (1) in Fig. 3.16. (ii) *Protein structure hypothesis*: Any real mature protein with appropriate three-dimensional structure have not been proposed by the protein structure hypothesis. Therefore, the hypothesis is incompatible with the condition (2) in Fig. 3.16 that even the most primitive one must have similar features to modern one. (iii) *Short peptide (motif) theory*: It would be impossible to create any protein with appropriate three-dimensional structure by assembling short peptides produced through random process with some exceptions (Antonkine et al. 2009; Gibney et al. 2000). Therefore, the short peptide (motif) theory is incompatible with the condition (1) in Fig. 3.16.

Thus, it would be a natural consequence that other researchers could not reach a correct conclusion for the origin of protein-II (Fig. 3.17), because they have not noticed the protein 0th-order structure, which is indispensable to understand the origins of protein-I and II.

On the contrary, immature but water-soluble globular [GADV]-proteins with some flexibility, which were produced by random polymerization of [GADV]-amino acids in the protein 0th-order structure, satisfy all the three conditions indicated in the Fig. 3.16. That is, (1) immature [GADV]-proteins, which were produced by random joining of [GADV]-amino acids, (2) could mature to modern proteins. Furthermore, maturation process from [GADV]-protein to modern protein can be also explained by coevolution with the genetic code from GNC code to the universal genetic code through SNS code, as supposed by GNC-SNS primitive genetic code hypothesis, which I have proposed (Fig. 3.18) (Ikehara et al. 2002).

A small modern protein is usually composed of 20 kinds of about 100 amino acids. Therefore, influence of one amino acid replacement in the protein becomes small by receiving a kind of buffer action of one of 100 amino acids. This makes it possible to evolve easily from an immature protein to a mature protein step by step through repeated amino acid replacements. Not only the case, that makes it possible to transform from one mature protein to another homologous mature protein, which was produced according to gene duplication theory (Ohno 1970) and also to generate various proteins through amino acid replacements induced by base substitutions or through creation of EntNew proteins from pan-GC-NSF(a), so that diverse organisms can inhabit in various extreme environments on the Earth (Fig. 3.19).

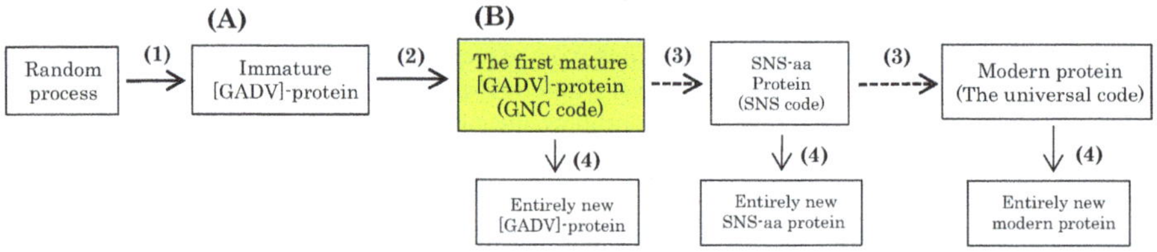

Fig. 3.18 The origin and evolution of protein-I and II (protein 0th-order hypothesis). (**a**) An immature [GADV]-protein was produced by random joining of [GADV]-amino acids ((1) Random process). (**b**) The first mature [GADV]-protein was created accompanied by creation of the first ds-$(GNC)_n$ RNA ((2) Essentially random process). (3) Thereafter, the formation process of a mature [GADV]-protein encoded by $(GNC)_n$ gene could be succeeded to modern protein encoded by $(NNN)_n$ gene through SNS-aa protein encoded by $(SNS)_n$ gene along the progression of coevolution of protein with gene. (4) During the evolution, EntNew proteins could be produced from the respective pan-GC-NSF(a)s, when necessary

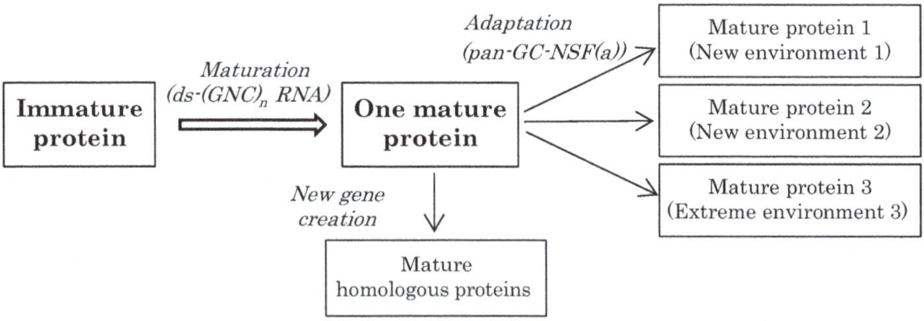

Fig. 3.19 Proteins are usually composed of amino acids more than one hundred. This could diminish influence to structure change of the protein induced by one amino acid replacement, resulting in generation of various kinds of proteins from one protein through cumulative amino acid replacements. Furthermore, creation of various EntNew proteins from pan-GC-NSF(a) has enabled for organism to adapt to various environments

6 Conclusion

One of the greatest concerns in life science fields would be to understand the mechanism how a splendid protein like a precision polymer machine could be created. Nowadays, it is well known that such a protein has been produced by gene expression or owing to the existence of gene. However, it has not been understood well thus far, how the first primitive protein was created in the absence of gene and also how such splendid EntNew mature proteins were and have been generated, because the riddle has not been solved how genes encoding such proteins could be created. On the other hand, it is also well known that there is a "chicken-egg" relationship between gene and protein, because protein cannot be produced in the absence of gene and genetic information cannot be expressed without protein. This also has made it difficult to understand the origin of protein until now.

On the contrary, I noticed that protein 0th-order structure, in which an immature but water-soluble globular protein with some flexibility could be produced by random joining of [GADV]-amino acids or in the absence of gene and that, therefore, the first, not a mature but an immature protein could be produced independently of gene (Ikehara 2002, 2005, 2009). Thus, the mechanism, how protein can be produced in the absence of gene, could be understood and simultaneously I was released from the curse of "chicken-egg" relationship between gene and protein. I have now considered that all EntNew modern proteins were and have been created through maturation of the respective immature proteins, which were produced by expression of pan-GC-NSF(a) on antisense strand of GC-rich genes or $(GNC)_n$ gene, $(SNS)_n$ gene and modern GC-rich gene, anytime if necessary.

My rough picture of protein, which is viewed through protein 0th-order structure hypothesis, is as follows.

Diverse proteins could be created through selection of a favorable structure from the extraordinary large structure space of immature proteins with flexible structure, which is induced upon wobbling of surface amino acids. The extraordinary structure space is of course supported by the extraordinary large amino acid sequence space.

Successively, one immature protein selected out from the extraordinary diverse immature proteins could be matured and optimized to a splendid mature protein.

Recently, many graphics of modern proteins with an amazing three-dimensional structure can be seen in various scientific journals and even in textbooks for students. We now look at the pictures of modern proteins, which have been optimized to the utmost limit (Fig. 3.1).

The reasons, why such beautiful proteins could be created, are of course basically based on 20 natural amino acids. Then, it is explained "What are amino acids?"

General Features of Amino Acids

Why are or must amino acids be used as main organic compounds in organisms living on the present Earth?

The reasons are because amino acids have the following quite outstanding features for forming versatile proteins.

(1) Amino acids are quite simple organic compounds, of which core structure is composed of only five atoms as C_2O_2N, if side chains and hydrogen atoms are ignored.
(2) Nevertheless, amino acids have both positive ($-NH_3^+$) and negative ($-CO_2^-$) charges in the molecule.
(3) In addition, free rotation around peptide bond (-CO-NH-), which is formed between two amino acids, is strictly restricted in a plane to make it possible to form easily regular structures as α-helix and β-sheet.

Therefore, it can be concluded that amino acids are generally quite simple organic compounds, which could be synthesized easily with prebiotic means on the primitive Earth and are excellent organic compounds, which can be used in proteins exhibiting admirable functions in organisms, as it would be probably impossible to find out other organic compounds possessing such amazing features. Then, next it is explained "What is protein?"

General Features of Protein Viewed from the Origin of Protein
It can be stated that protein is working polymers, which can exhibit a wide range of versatile functions according to a combination of amino acids in the respective polypeptide chains, although relatively simple amino acids are used for the polypeptide synthesis. On the other hand, gene only plays a role in carrying information for the polypeptide synthesis.

Such proteins with amazing functions have been always created from an immature and incomplete proteins with some flexibility, which were produced by direct joining of [GADV]-amino acids on the primitive Earth or through expression of pan-GC-NSF(a) or nonstop frame on antisense strand of GC-rich gene, as $(GNC)_n$ gene, $(SNS)_n$ gene and modern GC-rich gene after the first $(GNC)_n$-gene was created. That is, a mature protein with rather rigid structure was and has been created from an immature protein with some flexibility ever since after the first ds-$(GNC)_n$ RNA was created. In other words, every EntNew mature protein has been created through maturation process from the respective immature proteins produced under protein 0th-order structure after establishment of the genetic system. I firmly believe that there is only the way for creation of a mature protein, which is triggered by production of an immature protein with some flexibility.

References

Antonkine ML, Koay MS, Epel B et al (2009) Synthesis and characterization of de novo designed peptides modelling the binding sites of [4Fe-4S] clusters in photosystem I. Biochim Biophys Acta 1787:995–1008

Berg JM, Tymoczko JL, Stryer L (2002) Biochemistry, 5th edn. WH Freeman and Company, New York

Biscans A (2018) Exploring the emergence of RNA nucleosides and nucleotides on the early earth. Life 8:57

Bonner WA (1991) The origin and amplification of biomolecular chirality. Orig Life Evol Biosph 21:59–111
Breslow R, Levine MS (2006) Amplification of enantiomeric concentrations under credible prebiotic conditions. Proc Natl Acad Sci U S A 103:12979–12980
Caspar DL, Klug A (1962) Physical principles in the construction of regular viruses. Cold Spring Harb Symp Quant Biol 27:1–24
Cepelewicz J (2017) Life's first molecule was protein, not RNA. New model suggests (Dill KA, Guseva E, Zuckermann R). Quanta Magazine on November 12, 2017
Cleaves HJ, Chalmers JH, Lazcano A et al (2008) A reassessment of prebiotic organic synthesis in neutral planetary atmosphere. Orig Life Evol Biosph 38:105–115
Crick FH (1958) On protein synthesis. Symp Soc Exp Biol 12:138–163
Crick FH (1970) Central dogma of molecular biology. Nature 227:561–563
Dayhoff MO (1965) Computer aids to protein sequence determination. J Theor Biol 8:97–112
Dill KA (1990) Dominant forces in protein folding. Biochemistry 29:7133–7155
Fox SW, Dose K (1977) Fox JL (ed) A series of textbook, Biology, vol 2. Molecular evolution and the origin of life (revised edition)
Fox SW, Jungck JR, Nakashima T (1974) From protenoid microsphere to contemporary cell: formation of internucleotide and peptide bonds by protenoid particls. Orig Life 5:227–237
Furukawa Y, Chikaraishi Y, Ohkouchi N et al (2019) Extraterrestrial ribose and other sugars in primitive meteorites. Proc Natl Acad Sci 116:24440–24445
Gibney BR, Isogai Y, Rabanal F et al (2000) Self-assembly of heme A and heme B in a designed four-helix bundle: implications for a cytochrome c oxidase maquette. Biochemistry 39:11041–11049
Guseva E, Zuckermann RN, Dill KA (2017) Foldamer hypothesis for the growth and sequence differentiation of prebiotic polymers. Proc Natl Acad Sci U S A 114:E7460–E7468
Higgs PG (2009) A four-column theory for the origin of the genetic code: tracing the evolutionary pathways that gave rise to an optimized code. Biol Direct 24:4–16
Ikehara K (2002) Origins of gene, genetic code, protein and life: comprehensive view of life system from a GNC-SNS primitive genetic code hypothesis. J Biosci 27:165–186
Ikehara K (2005) Possible steps to the emergence of life: the [GADV]-protein world hypothesis. Chem Rec 5:107–118
Ikehara K (2009) Pseudo-replication of [GADV]-proteins and origin of life. Int J Mol Sci 10:1525–1537
Ikehara K (2014) Protein ordered sequences are formed by random joining of amino acids in protein 0th-order structure, followed by evolutionary process. Orig Life Evol Biosph 44:279–281
Ikehara K (2019) The origin of tRNA deduced from *Pseudomonas aeruginosa 5′* anticodon-stem sequence: anticodon stemloop hypothesis. Orig Life Evol Biosph 49:61–75
Ikehara K (under review) The origin of metabolism and GADV hypothesis on the origin of life. Smoukov SK, Seckbach J, Gordon R (eds) Conflicting Models for the Origin of Life [Volume 1 in the series astrobiology perspectives on life of the universe, Eds. Gordon R, Seckbach J]. Wiley-Scrivener, Beverly, Massachusetts
Ikehara K, Amada F, Yoshida S et al (1996) A possible origin of newly-born bacterial genes: significance of GC-rich nonstop frame on antisense strand. Nucl Acids Res 24:4249–4255
Ikehara K, Omori Y, Arai R et al (2002) A novel theory on the origin of the genetic code: a GNC-SNS hypothesis. J Mol Evol 54:530–538
Koga T, Naraoka H (2017) A new family of extraterrestrial amino acids in the Murchison meteorite. Sci Rep 7:636
Kojo K (2010) Origin of homochirality of amino acids in the biosphere. Symmetry 2:1022–1032
Koshland DE Jr (1995) The key–lock theory and the induced fit theory. Chemie, Angewandte
Miller SL, Orgel LE (1974) The origins of life on the earth. Prentice-Hall, Englewood Cliffs
Oba T, Fukushima J, Maruyama M et al (2005) Catalytic activities of [GADV]-peptides: formation and establishment of [GADV]-protein world for the emergence of life. Orig Life Evol Biosph 35:447–460

Ohno S (1970) Evolution by gene duplication. Springer, Heiderberg
Patel BH, Percivalle C, Ritson DJ et al (2015) Common origins of RNA, protein and lipid precursors in a cyanosulfidic protometabolism. Nat Chem 7:301–307
Pizzarello S (2006) The chemistry of life's origin: a carbonaceous meteorite perspective. Acc Chem Res 39:231–237
Pizzarello S, Weber AL (2004) Prebiotic amino acids as asymmetric catalysis. Science 303:1151
Romero MLR, Rabin A, Tawfik DS (2016) Functional proteins from short peptides: Dayhoff's hypothesis turns 50. Angew Chem Int Ed 55:15966–15971
Van der Gulik P, Massar S, Gilis D et al (2009) The first peptides: the evolutionary transition between prebiotic amino acids and early proteins. J Theor Biol 261:531–539

Chapter 4
The Origin of Cell Structure

Abstract It has been considered by many researchers at this time point that cell originated from vesicles or micelles, which were formed with amphiphiles like fatty acids, monoglycerides and phospholipids. The experimental results were published by Deamer, showing that vesicles could be formed by amphiphilic lipids, which could be extracted from Murchison meteorite (Deamer DW, Pashley RM, Ori Life Evol Bioshp 19:21–38, 1989). However, even protocell should be not a simple bag but a functional bag, of which membrane can pass through various small organic compounds. If the small organic compounds, which accumulated around the protocell, could not pass through the membrane, the protocell could not survive for a long time. On the contrary, I have proposed in this Chapter the idea, suggesting that the first protocell must be a [GADV]-protenoid microsphere, into which small chemical compounds as water, glyoxylate, glycelaldehyde, pyruvate and so on can be incorporated. Furthermore, I consider that the protocell without any gene could proliferate and evolve owing to individuality of the protocell and catalytic activities exhibited by [GADV]-proteins. The idea or [GADV]-protenoid microsphere hypothesis, assuming that [GADV]-protenoid microsphere was protocell, is of course similar to Oparin's idea, coacervate theory ((Oparin AI, The origin of life on earth. Academic, New York, 1957), and especially to Fox's idea, protenoid microsphere theory (Fox SW, Dose K, Molecular evolution and the origin of life. Marcel Dekker Inc., New York, 1977). It is considered that the first genuine life emerged when the [GADV]-microsphere acquired double-stranded $(GNC)_n$ genes, of which formation was triggered by creation of anticodon stem-loop tRNA through random process.

Keywords Origin of cell structure · Cell structure · Amphiphile membrane theory · Protenoid microsphere theory · [GADV]-microsphere hypothesis · [GADV]-protein world · Protocell

Sydney W. Fox, who had studied on formation of protenoid microsphere for a long time, described in their paper as follows (Fox et al. 1974).

> *Among the contributors whose ideas can be interpreted as consonant with that of a cell initiating protein and nucleic acid synthesis and Darwinian selection have been Wald*

K. Ikehara, *Towards Revealing the Origin of Life*,
https://doi.org/10.1007/978-3-030-71087-3_4

(1954), Van Niel (1956), Oparin (1957), Lederberg (1959), Ehrensvard (1962), and Prosser (1970). The integrated view from their hypotheses and from our experimental results is expressed concisely by the sequence: protoprotein --- heterotrophic protocell --- macromolecule-synthesizing cell.

Of course, it seems to me that the Fox's idea on the origin of protocell described above is plausible and natural. However, the protein or amino acid-centered excellent idea was denied by many researchers mainly working in RNA world hypothesis, which is based on gene-centered idea, and was buried. The reasons would be because many researchers consider that the emergence of life should not be triggered by proteins and protenoid microspheres, even if those could be produced by prebiotic means and those could exhibit some catalytic activities as described by Fox et al. (1974).

However, I feel now that the steps to the emergence of life should proceed according to the sequence: *protoprotein* ([GADV]-protein) --- *heterotrophic protocell* ([GADV]-protenoid microsphere) --- *macromolecule-synthesizing cell* (the first cell synthesizing [GADV]-proteins and $(GNC)_n$ RNA genes), as Fox et al. expected. Then, I explain my idea about the way how the first protocell was created through random processes on the primitive Earth as referring basic features of modern cell in this chapter (Figs. 4.1 and 4.2).

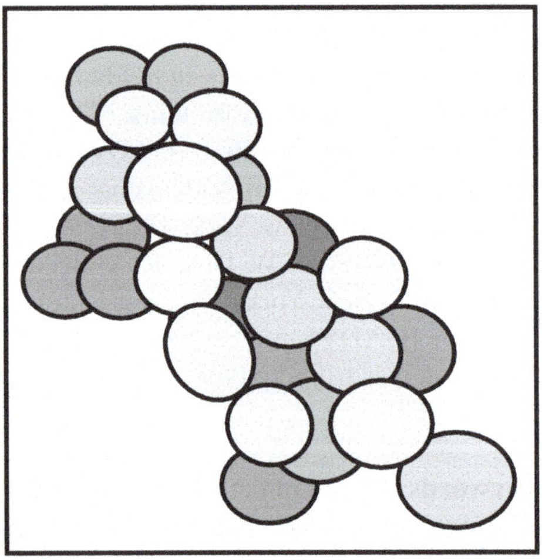

Fig. 4.1 A drawing traced an scanning electron microscope image of *Staphylococcus aureus* (Baba et al. 2002). As like the microorganism, why was the first cell structure or [GADV]-microsphere globular? *S. aureus*, which is living with about 2600 genes, is small but is one of splendid organisms surrounded by cell membrane. How could various amazing organisms emerge on this Earth?

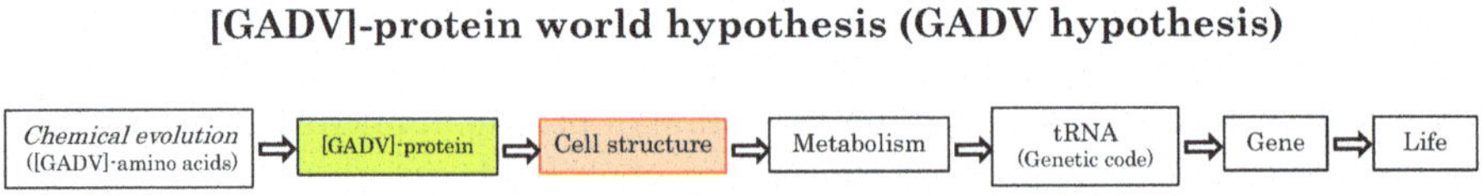

Fig. 4.2 In this Chapter, the origin of cell structure is discussed (red box). The most primitive cell structure was constructed with immature [GADV]-proteins, which were produced by random joining of [GADV]-amino acids or, for example, by repeated wet-drying cycles of [GADV]-amino acids (yellow box)

Introduction
In this Chapter, I first discuss what happened during chemical evolution before the emergence of life, because it is considered that the first cell structure or [GADV]-microsphere was formed using [GADV]-amino acids, which were synthesized with prebiotic means and accumulated on the primitive Earth. Furthermore, it is presumed that a kind of evolution of [GADV]-microsphere could occur among the protocells or actually among the protenoid microspheres without any genetic function.

1 Towards Solving the Riddle of the Origin of Cell Structure

1.1 Necessary Matters for Exploration of the Origin of Cell Structure

Before considering the origin of modern cell itself, the conditions necessary to explore the origin of modern cell are enumerated as follows.

1. It must be clarified as understanding the features of modern cell, from what kind of primitive cell structure the contemporary one originated.
2. Successively, evolutionary process from the primitive protocell, which could be created by random reactions on the primitive Earth, to modern cell must be explained reasonably.
3. If it is assumed that the contemporary cells or organisms originated from a primitive one with different features from modern cells, as a matter of course, conversion process from the primitive protocell with different features to the most primitive one with similar features to modern organisms must be reasonably explained, either directly or indirectly.

1.2 General Structural Features of Modern Cell (Cell Structure)

1. Cell or cell structure is surrounded by a membrane.
2. Contemporary cell membrane, which is drawn as a fluid mosaic model, contains both membrane proteins and phospholipids.
3. Cell is sufficiently large to be able to enclose a large number of proteins and organic compounds, so that the parental cell can transmit various functions to progeny cells after cell division.
4. Cell must be sufficiently stable to maintain the extremely large structure compared to proteins and polynucleotides.

1.3 General Functional Features of Modern Cell (Cell Structure)

1. Cell is not a simple bag but a functional bag, which equips with functions to secure enough permeability of a low molecular weight chemical compounds and inversely to retain polymers as proteins and nucleic acids, which were synthesized with the chemical compounds incorporated into the cell.
2. Cell proliferates as transmitting various functions to progeny cells upon cell division or self-reproduction. Therefore, proteins with a high function are synthesized in the cell at a sufficiently high rate to compensate a deficiency upon cell division, so that cells can continuously proliferate.
3. Cell evolves through selection among progeny cells.

2 The Ideas on the Origin of Cell Structure Advocated by Other Researchers

It is one of the key points leading to solve the origin of life to understand the process, how a protocell enclosing various chemical compounds could be made on the primitive Earth and triggered to evolve to modern cell, even if the primitive cell-like structure had not held on genetic materials. Taking this also into consideration, it is discussed in this Section, whether or not the ideas proposed by other researchers are valid to the origin of the most primitive cell structure.

2.1 Amphiphile Membrane Theory

Amphiphile molecules, which are simpler than phospholipids used by many modern organisms, can spontaneously assemble to form stable membranous vesicles. For instance, fatty acids like oleic acid readily form vesicles. In 1985, Deamer extended this observation to the amphiphilic compounds, which were extracted from carbonaceous Murchison meteorite, to show that they can be also assembled into microscopic vesicles (Deamer and Pashley 1989). The discovery that two-layered vesicles can be formed with the amphiphiles extracted from the Murchison meteorite, which came from space, strengthened the theory on the origin of cell membrane. Thus, many researchers believe that the first cell membrane was formed with the amphiphiles.

Strengths of Amphiphile Membrane Theory One of characteristics of life is individuality, which is realized by cell membrane. The amphiphile membrane theory

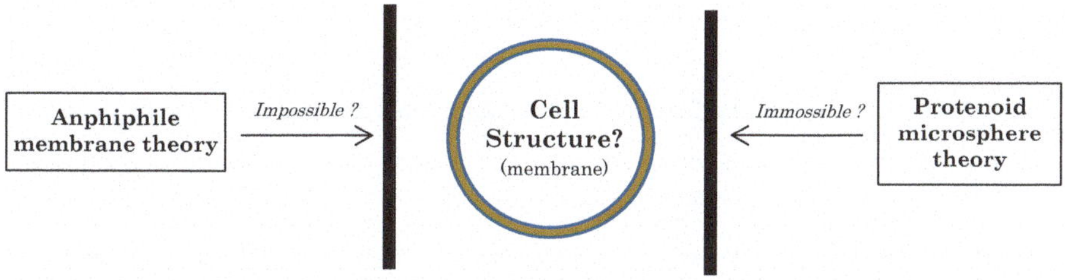

Fig. 4.3 Protocell could not be created as suggested by amphiphile membrane theory, because the membrane is catalytically inactive and could not support proliferation of the protocell. Furthermore, Lys-based protenoid microspheres including Leu, Met and so on as advocated by Fox and Dose (1977) were not realized on the primitive Earth, because some complex amino acids including Lys, which they used in their experiments, should not be accumulated at a large amount on the primitive Earth

has been proposed as focusing on the way how the first life acquired what kind of membrane. It has been confirmed by experiments that a possible protocell membrane can be formed with amphiphiles extracted from Murchison meteorite (Deamer and Pashley 1989). The fact is the greatest strength of the theory. Amphiphile membrane in the theory corresponds to [GADV]-protein membrane in GADV hypothesis.

Weaknesses of Amphiphile Membrane Theory However, the vesicles made with amphiphiles are simple bags, which have not any proteinaceous enzyme (Fig. 4.3). Therefore, the vesicles could not exhibit cell like activities. Of course, many researchers endorsing the amphiphile membrane theory may object against the idea, that the vesicles made with amphiphiles are simple bags, by arguing that the vesicles could incorporate proteinaceous catalysts into the inside. If so, proteins should be contained in the primitive amphiphile membrane, meaning that the primitive membrane was complex membrane composed of amphiphiles and proteins like modern membrane from the beginning but not pure amphiphile membrane.

In modern metabolic pathway, syntheses of fatty acids and lipids are carried out through the pathway similar to from oxaloacetate to succinic acid on reverse-TCA cycle. Taking coevolution theory (Wong 1975; Di Giulio 2008) into consideration, it would be more reasonable to consider that lipids were synthesized after the partial TCA cycle was completed. Furthermore, fatty acids are synthesized through a complex metabolic pathway starting from incorporation of acetyl-CoA into malonyl-CoA to produce butylyl-CoA. These facts would support the [GADV]-protein membrane hypothesis, assuming that [GADV]-protein membrane was used in the most primitive protocell earlier than the use of amphiphiles and lipids in the membrane.

2.2 Protenoid Microsphere Theory

The study on microsphere by Fox is well introduced in a recent Z. Hua's paper (2018):

> *When a mixture of various amino acids was heated, polymerization occurred among amino acids and a protein-like hyperpolymer with a molecular weight as high as approximately 8,000 to 20,000 Dalton was produced. When these proteinoids were dissolved in water, they would automatically aggregate into a microspheric multi-molecular system, which he termed a microsphere. Microspheres resemble bacteria, have a double-layer boundary, and internal structure. They proliferated themselves through budding and division, and had weak catalysis. Fox believed that such microspheres were the models for primitive bacteria.*

Strengths of Protenoid Microsphere Theory Many significant experimental results, considerations regarding formation of microspheres and various catalytic reactions with microspheres were provided by Fox et al. (1974). In addition, Fox had well understood that modern cell originated from protenoid microsphere. However, unfortunately, it seems to me that their eminent theory was swept away by RNA world hypothesis (Gilbert 1986) and gradually disappeared, as the RNA world hypothesis, which stands on the side of gene/replicator-early theory but not protein/metabolism early theory, is accepted by many researchers. Protenoid membrane proposed by Fox corresponds to [GADV]-protenoid membrane in the GADV hypothesis.

Weaknesses of Protenoid Microsphere Theory Acidic amino acids as Asp and Glu, basic amino acids as Lys and hydrophobic amino acids were mainly used to prepare protenoid microspheres by Fox's group (Fox and Dose 1977). Therefore, the importance of [GADV]-protenoid microsphere about the origin of life might not be well recognized by Fox. Furthermore, in the protenoid microsphere hypothesis, the origins of tRNA, genetic code and gene, which are intimately related to the origin of life, had not been discussed. Therefore, Fox could not clearly point out the evolutionary steps from the protenoid microsphere to the emergence of life (Fig. 4.3).

3 My Idea About the Origin of Modern Cell

3.1 For What Purpose Was the First Cell Structure Created?

Was the first cell structure formed for generating the first genuine life or toward a predetermined purpose? Or accidental formation of cell structure simply resulted in the emergence of life? In the following Sections, it is discussed with such problems in mind, how the first cell structure was created with what kind of amino acids.

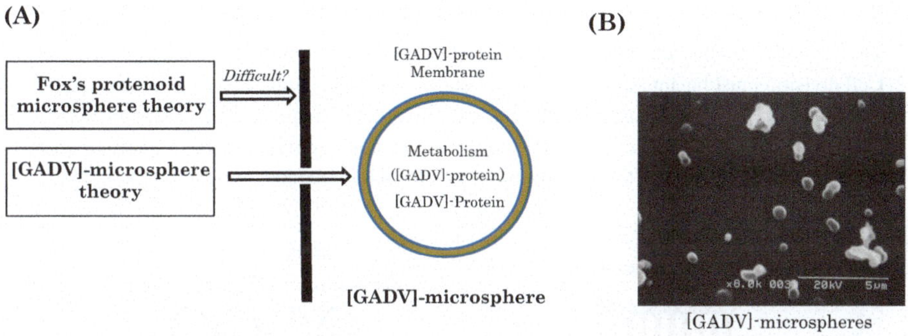

Fig. 4.4 (**a**) It may be difficult to explain the establishment process of the fundamental life system with Fox's protenoid microsphere, because Fox's group used rather complex amino acids as Lys, Leu and Met too (Fox and Dose 1977). On the other hand, [GADV]-protenoid microsphere could overcome the high wall, because the microspheres were composed of only simple [GADV]-amino acids, which could be produced with prebiotic means. (**b**) Scanning electron microscope image of [GADV]-protenoid microspheres, which were produced by repeated wet-drying processes of [GADV]-amino acids (Oba et al. 2005)

3.2 [GADV]-Microsphere Hypothesis

3.2.1 Formation of [GADV]-Microsphere

[GADV]-microspheres could be easily formed in any present laboratory by repeated wet-drying processes of [GADV]-amino acids, 30 times at 90 °C, according to the methods described in the paper published by Shuwannachot and Rode (1998). Therefore, [GADV]-microspheres could be also easily formed in repeated wet-drying cycles of [GADV]-amino acids in depressions of rocks on seashore, as [GADV]-amino acids were produced and accumulated at large amounts on the primitive Earth. Hence, it is supposed that [GADV]-microspheres were the first protocells (Oba et al. 2005) (Fig. 4.4). Of course, the [GADV]-protenoid membrane might contain amphiphles from the very beginning, as fatty acids and lipids, which were synthesized with prebiotic means and accumulated around the microsphere, although such amphiphiles could only play an auxiliary role in the membrane. Thus, [GADV]-protein world would be formed in the [GADV]-microsphere and various kinds of metabolic reactions could be carried out with immature [GADV]-proteins in the microsphere, as described in more detail in Chap. 5.

3.2.2 Properties of [GADV]-Microsphere

1. *Construction of cell membrane*: Cell membrane of the first protocell should be composed of chemical materials, which could be supplied from the environments and had accumulated with prebiotic means on the primitive Earth until the protocell was formed (Table 4.1 (1)). Taking this into consideration, [GADV]-proteins,

Table 4.1 Necessary conditions for constructing cell structure (membrane)

1. The first cell structure must be constructed with materials accumulated on the primitive Earth
2. Cell division must be autonomously carried out by synthetic ability of cell structure, itself.
3. Basic metabolic system must be formed in the cell structure to proliferate continuously.
4. Cell membrane of the first cell structure must be not amphiphile membrane but proteinaceous membrane to passively but efficiently transport chemical compounds through the membrane.

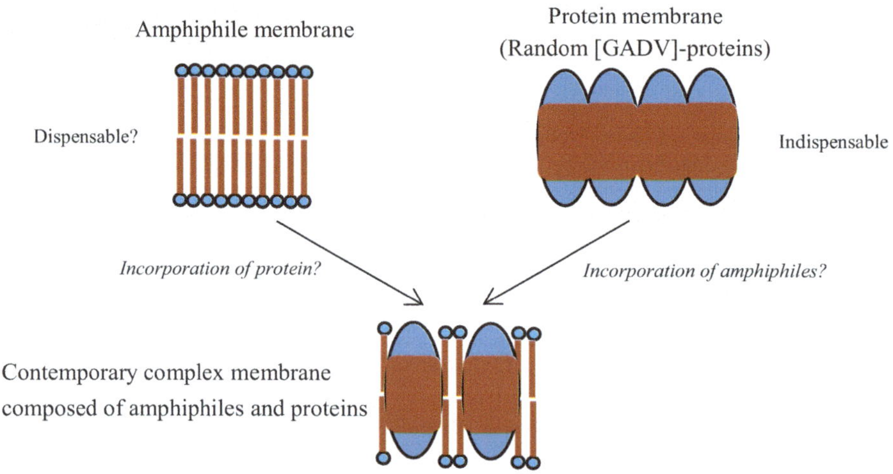

Fig. 4.5 Contemporary complex cell membrane is composed of both amphiphiles and proteins. Therefore, proteins should be incorporated into amphiphile membrane one day, according to amphiphile membrane theory on the origin of cell membrane. On the other hand, according to the [GADV]-protein membrane hypothesis, which I have proposed, amphiphiles as fatty acids and lipids should be captured into the protein membrane one day

actually aggregates of [GADV]-peptides, should be used for construction of the first membrane, because [GADV]-proteins have both cell membrane formability and catalytic functions, although amphiphiles or lipids, which were accumulated with prebiotic means around the cell-like structure, should be also incorporated into [GADV]-protein membrane as like contaminants from the beginning. However, it is supposed that sufficient amounts of lipids were incorporated into the membrane after lipid metabolism could be developed in the [GADV]-microsphere (Fig. 4.5). Note that it was not the necessary condition for construction of the protocell, that gene as ds-RNA or ds-DNA must be contained in the protocell (Table 4.1).

2. *Why was the [GADV]-protenoid membrane used for the first protocell?*: The reason is because [GADV]-protenoid membrane was optimal as the chemical material, which can be used for formation of cell-like structure, as described below.

1. [GADV]-protein membrane was constructed with [GADV]-amino acids, which were synthesized with prebiotic means and accumulated at a large amounts on the primitive Earth.

2. [GADV]-protenoid microspheres could be produced with a simple method or repeated wet-drying processes.
3. [GADV]-proteins with various catalytic activities could be incorporated into the cell-like structure.
4. [GADV]-proteins synthesized through pseudo-replication of [GADV]-protein in the microsphere could raise the internal osmotic pressure to induce growth and division of the microsphere.

Therefor, it can be considered that the first cell structure was not formed toward a predetermined purpose that the first genuine life must be generated, but was accidentally formed, simply resulted in the emergence of life, although it can be also considered that the way to the formation of [GADV]-microsphere was prepared by something in advance.

3. Formation of [GADV]-protein world: [GADV]-protein world was formed in [GADV]-microspheres, so that the initial metabolic system could be established to efficiently produce [GADV]-amino acids, [GADV]-peptides, nucleotides and RNA in the [GADV]-protein world (Chap. 5: Sect. 3.1). In addition, the chemical materials, which were used to construct the protocell membrane, must be continuously supplied through the initial metabolic system, because it is easily supposed that the chemical materials would be depleted from the protocell accompanied by using them for construction of the membrane of its descendants.

4. Syntheses of [GADV]-amino acids and nucleotides in [GADV]-microsphere: Various organic compounds, as glyoxylate (Go), glyceraldehyde (Ga) and pyruvate (Pyr) etc. with a low molecular weight, were synthesized with prebiotic means and accumulated at large amounts on the primitive Earth, and could be incorporated into [GADV]-protenoid microsphere, which were newly formed after the protocell division (Fox and Dose 1977). Then, [GADV]-amino acids and nucleotides were synthesized with the three GoGaPyr organic compounds as substrates (Chap. 5: Sect. 3.1).

5. Why is [GADV]-microsphere literally sphere?: The reason, why [GADV]-microsphere is sphere, would be probably because protocell or microsphere could take in passively organic compounds with a low molecular weight and water through the [GADV]-protenoid membrane. [GADV]-microsphere could swell after addition of water through rainfall on the primitive Earth to heat-dried microspheres (Fig. 4.6). This caused to make the literally spherical [GADV]-microspheres. This clearly indicates that microsphere can incorporate passively chemical compounds with a low molecular weight, which accumulated with prebiotic means, because the inside of protenoid microsphere was placed under a high osmotic pressure through [GADV]-peptide synthesis. Then, [GADV]-peptides and proteins synthesized by immature [GADV]-proteins could be inserted into everywhere of [GADV]-protenoid membrane and [GADV]-microspheres could be divided (Fig. 4.6) (Hua 2018). It could be stated that the first but preliminary life without any genetic system might emerge at the time point when [GADV]-microsphere divided upon [GADV]-peptide synthesis under the first primitive metabolism, because it is expected that Darwin evolution also began upon the division of

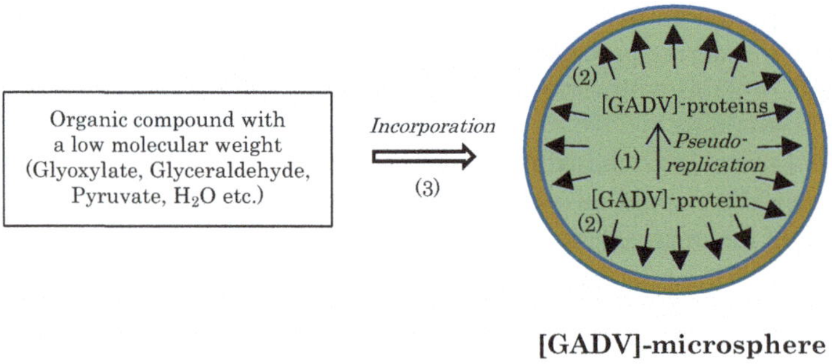

Fig. 4.6 Water and organic compounds with a low molecular weight could be incorporated into the [GADV]-protenoid microsphere, because the inside was maintained at a high osmotic pressure (1) by pseudo-replication of [GADV]-protein. It is considered that the spherical structure of the [GADV]-microsphere was formed by the high internal osmotic pressure, and (2) by insertion of [GADV]-peptides/proteins into everywhere of the membrane from the inside. Light green inside of the [GADV]-microsphere shows the high osmotic pressure

[GADV]-microspheres, so that the first primitive metabolism could be developed through selection of [GADV]-protenoid microspheres with a higher proliferation ability than others (Sect. 3.2.2).

6. *Proliferation of the protocell*: The protocell or microsphere must be divided autonomously with the ability of the protocell itself synthesizing polymeric components as [GADV]-proteins in the protocell (Hua 2018). Inversely stating that, it would be impossible to divide the protocell continuously only incorporation of low molecular weight chemical compounds accumulated around the protocell. Therefore, a basic metabolic system, in which various organic compounds especially polymers as [GADV]-peptides could be produced with proteinaceous catalysts, would be formed in the protocell. This simultaneously means that the membrane of the protocell should transport passively various organic compounds to use as substrates for syntheses of the polymeric components of the protocell. Note that similar idea on the self-reproduction of vesicle but not [GADV]-microsphere is given in Figure 4.4 of Luisi's book (Luisi (2016)).

On the other hand, it is considered that driving force of cell division of extant unicellular microorganisms is also a high internal osmotic pressure, which is caused by accumulation of biopolymers as proteins in the inside. Therefore, the protocell could be divided and proliferated, if a low molecular weight organic compounds as glyoxylate and pyruvate were incorporated into the protocell and polymers as [GADV]-petides could be constantly produced in the protocell using [GADV]-amino acids as substrates. [GADV]-amino acids could be also synthesized by reactions catalyzing amination of glyoxylate and pyruvate in the protocell. Hence, a low molecular weight organic compounds as glyoxylate and pyruvate could be passively incorporated into the protocell upon the increasing of internal osmotic pressure, which was caused by accumulation of [GADV]-peptides. Such reaction cycles should be easily induced, because that is a simple physicochemical phenomenon. In addition, it is also easily supposed that [GADV]-proteins as aggregates of

[GADV]-peptides could synthesize [GADV]-peptides, because it is known that even glycine peptides as Gly-Gly and Gly-Gly-Gly catalyze peptide bond formation (Gorlero et al. 2009).

7. *Effective concentration of organic compounds*: A favorable situation for the emergence of the first life could be generated by two effects for concentration of organic compounds as described below.

1. Organic compounds with a small molecular weight could be effectively introduced into [GADV]-microspheres. By this, many small but useful organic compounds as [GADV]-amino acids for the emergence of life could be effectively concentrated from the environment around the microspheres into the inside of the microspheres.
2. Furthermore, an environment containing various organic compounds at a high concentration could be generated by precipitation of a large number of inanimate [GADV]-microspheres upon drying up them followed by degradation. With that too, the environment, which was useful for the first life to emerge, could be formed for example in depressions of rocks on the primitive Earth.

8. *Evolution of [GADV]-microsphere without gene*: It is necessary to raise catalytic activities in the protocell without a genetic system as a whole in order to evolve step by step as propagating catalytic activities of a parental protocell to progenies during every division of the microspheres. Many persons may consider that it would be impossible for the microspheres to divide and proliferate. However, the pluripotency of immature [GADV]-proteins, which is described in Chap. 3: Sect. 4.7, would make it possible for the microspheres without any genetic system to evolve. The reason is because many catalytic activities of parental microspheres could be transferred to progeny microspheres because of the sufficiently high ability of pseudo-replication of immature but pluripotent [GADV]-proteins and a large amount of [GADV]-peptides accumulated in the parental microspheres (Ikehara 2009).

Furthermore, [GADV]-microsphere was sufficiently large to store [GADV]-proteins in the inside or [GADV]-protein world, in which various metabolic reactions could be carried out. A high division ability of the [GADV]-microsphere enabled to select [GADV]-microspheres with a higher division ability than others during growth and division of [GADV]-microspheres containing [GADV]-proteins with various amino acid sequences (Fig. 4.7). Of course, the microspheres without any genetic material could not use the memorizing ability of ds-RNA for amino acid replacements. However, the memorizing ability of the double-stranded (ds)-RNA could be substituted by a high proliferation ability of [GADV]-microspheres generating a range of probability distribution, although, of course, the efficiency was quite low as a gearing apparatus without ratchet, until genes could be created (Fig. 4.7).

9. *Two-types of cell structures*: Before consideration of the emergence of life, it must take notice of that there were two types of cell structures, one is a preliminary life or protocell without genetic material or RNA, and the other is a genuine life carrying the fundamental life system including ds-RNA genes. Then, I would like to discuss here about the two types of cell structures.

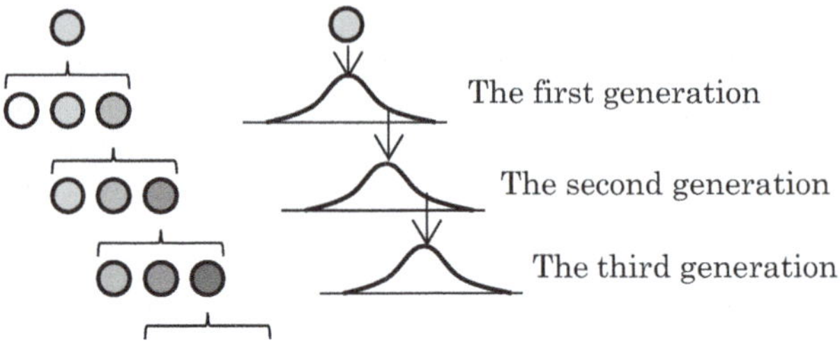

Fig. 4.7 Even [GADV]-protenoid microspheres without gene could evolve, because of a high proliferation ability of the microspheres and a wide probability distribution generated at each generation. A darker gray circle indicates a microsphere with a higher proliferation ability

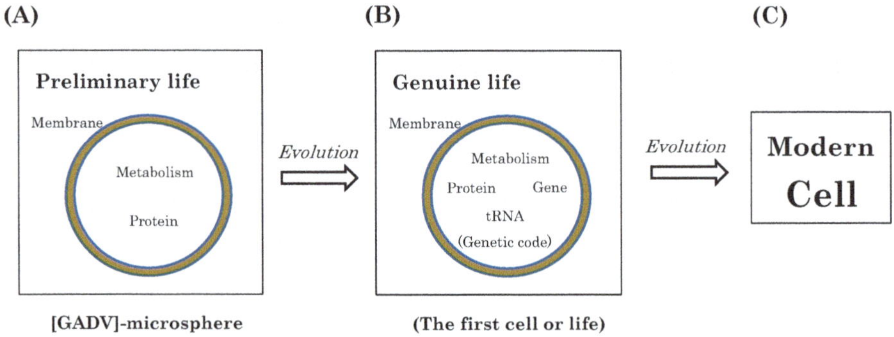

Fig. 4.8 How did [GADV]-microsphere evolve to the first cell and modern organism? (**a**) A preliminary life or a simple protocell was formed by aggregation of [GADV]-proteins. The protocell without any gene could proliferate and even evolve (see also Fig. 4.7). (**b**) The first genuine life, which evolved from the preliminary life without genetic system, emerged accompanied by acquisition of the fundamental life system composed of the six members. (**c**) Modern cells, to which descended from a preliminary life through a genuine life, inhabit under the essentially same fundamental life system composed of protein, cell structure, metabolism, tRNA, genetic code and gene, which were developed in the preliminary life and the genuine life, although the functions of the respective members have been highly sophisticated in the modern cells

The first cell structure: Preliminary life or protocell without gene: As described below, Fox and some other researchers already had noticed that life-like cell structure without any genetic system can be "self-reproduced" (see the sentences described by Fox et al. (1974) below). On the other hand, I also independently noticed during exploration of the origin of life through GADV hypothesis that [GADV]-protenoid microshpere without gene could divide, proliferate and even evolve (Fig. 4.8a) (Sect. 3.2.2).

> *The coacervate droplets of Oparin have demonstrated utility in the protection of cellular contents, in the promotion of intracellular reaction sequences, and favoring of a concentration of growing polymers. In 1967, we reported the evidence that protenoid microspheres are organized structures capable of 'self'-reproduction (Fox et al. 1967; Lehninger 1970; Szent Gyoergyi 1972).*

The protocell produced before the emergence of the genuine life, of course, did not possess any genetic material. Nevertheless, the protocell could proliferate and even evolve owing to selection among diverse protocells with different proliferation ability (Fig. 4.7). Formation of protocells as [GADV]-microspheres generated the possibility of selection and evolution. This also would be one of necessary matters for the emergence of life. Therefore, it might be possible to state that the time point, when [GADV]-protenoid microshpere was first formed, was the time when the first but preliminary life emerged (Fig. 4.8a).

The second cell structure: Genuine life or cell with gene: Genuine life can be self-replicated and proliferate under the genetic system. In addition, the genuine cell, of course, could evolve through selection among cells proliferating at a different division rate (Fig. 4.8b).

10. Improvement and development of protocell function made it possible to evolve to modern cell: As described above, even the protocell without genetic system could have an evolutionary ability. The individuality and the proliferation ability of the protocell enabled to evolve to modern life through the first genuine life (Fig. 4.8b). The first genuine life emerged after acquisition of ds-$(GNC)_n$ RNA/DNA genes. Therefore, the genuine cell could increase the number of genes/proteins and could develop the initial metabolism as increasing efficiently the number of kinds of tRNAs and codons encoding various amino acids like as a gearing apparatus having ratchets. Consequently, modern organisms are living as making full use of the six members constituting the fundamental life system, of which the respective functions have been optimized.

Strengths of [GADV]-Microsphere Hypothesis According to the necessary conditions for the origin of something, it could be concluded that [GADV]-microsphere is valid as the origin of modern cell, because both formation of [GADV]-microsphere through random process and evolutionary process from the protocell to modern cells can be reasonably explained based on GADV hypothesis as follows.

1. [GADV]-microsphere could be easily formed accompanied by production of [GADV]-proteins, actually aggregates of [GADV]-peptides, which were synthesized through random process or by repeated wet-drying cycles on the primitive Earth. Furthermore, an evolutionary process from the [GADV]-protein membrane to a modern complex membrane composed of proteins and amphiphiles can be also reasonably explained.
2. It was confirmed by experiments that [GADV]-proteins could play an important role in exhibiting some catalytic activities (Oba et al. 2005).
3. Formation of [GADV]-microspheres was confirmed with scanning electron microscope images obtained by Oba et al. (2005) (Fig. 4.4b). The first primitive protocell must be sphere, because [GADV]-microspheres could grow by random insertion of [GADV]-proteins into the protein membrane. Inversely, it would be difficult to construct a rod-shaped cell structure from the beginning, because formation of the structure must require a special growing point or a complex mechanism for membrane formation from the beginning.

4. Catalytic activities of the protenoid microspheres were also confirmed by experiments, which were carried out by Fox's group and Ikehara's group.

Weaknesses of [GADV]-Microsphere Hypothesis I have considered that evolution to modern cells started from a [GADV]-microsphere, which was formed by random joining of [GADV]-amino acids in protein 0th-order structure. However, simultaneously I have also considered that there is a great weakness in the [GADV]-microsphere hypothesis that several matters about the origin of cell structure have not been confirmed with experiments as described below.

1. Whether or not [GADV]-peptides can be really synthesized with [GADV]-peptides in [GADV]-microsphere, as expected by [GADV]-microsphere hypothesis.
2. Whether [GADV]-microsphere can be really divided and proliferate upon the [GADV]-peptide synthesis followed by increasing internal osmotic pressure of the microsphere or not, and so on.

4 Discussion

Two theories have been mainly proposed as the origin of cell structure, (1) amphiphile membrane theory and (2) protenoid membrane theory (Fig. 4.9). Then, consider which theories are valid for the origin of cell structure.

Then, I enumerate the conditions for suitable to the origin of cell structure below.

1. The original cell structure must be produced through random process.
2. The most primitive protocell must have features similar to the modern one.
3. If a protocell, which is dissimilar to the modern ones, is assumed, the dissimilar protocell must be transformed to the protocell with similar features to modern cell.

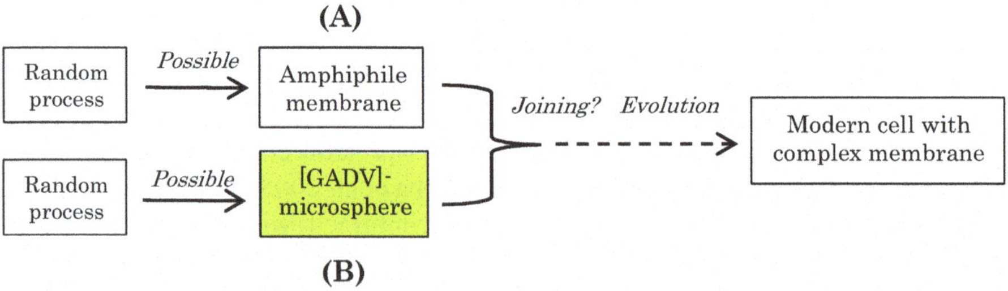

Fig. 4.9 (**a**) Ampjiphile membrane could be produced through random process. However, it would be a great weakness of the amphiphile theory that amphiphile membrane could not exhibit various catalytic activities necessary to proliferation. (**b**) Protenoid microsphere, especially [GADV]-microsphere, could be produced through random process or by repeated wet-drying cycles of [GADV]-amino acids. It is also confirmed by experiments that [GADV]-microsphere containing [GADV]-peptides have some catalytic activities including [GADV]-peptide synthetic activity. Therefore, it is supposed that [GADV]-microsphere could grow and proliferate

4. Even the most primitive protocell must have catalytic activities inducing cell division or proliferation.
5. In addition, the protocell assumed must be able to evolve to modern cell.

(1) Amphiphile membrane theory: It is supposed by the theory that the first membrane was composed of amphiphiles, as fatty acids, lipids etc. (Fig. 4.9). However, it is not concretely described in the theory about what kinds of catalytic activities amphiphile vesicles could express. Therefore, I have to state that the amphiphile theory on the origin of protocell does not satisfy the condition 4: because even the most primitive protocell must have some catalytic activities necessary to proliferate.
(2) The protenoid membrane theory proposed by Fox (Fox and Dose 1977) well satisfies all the conditions 1–5 described above (Fig. 4.9). However, Fox did not show the more primitive protenoid membrane as [GADV]-protenoid membrane, probably because Fox simply and unfortunately did not notice the significance of the [GADV]-protenoid membrane.

Therefore, I am honestly wondering especially about the reason why the "protenoid microsphere hypothesis", which was proposed by Fox, was ignored by many researchers and was buried, because protenoid microsphere had two indispensable abilities for protocell formation. One is for proteinaceous membrane formation and the other is catalytic activities including peptide synthesis leading to evolution to the modern cell.

On the contrary, I have proposed GADV hypothesis on the origin of life, assuming that life originated from [GADV]-protein world, which was formed by immature [GADV]-proteins. During the consideration of the GADV hypothesis, I noticed that the first protocell should be formed by [GADV]-proteins or should be [GADV]-protenoid microsphere as hypothesized by Fox. We, actually my students, tried to construct [GADV]-microsphere by repeated wet-drying processes about 15 years ago and confirmed formation of [GADV]-microsheres with scanning electron microscope image (Fig. 4.4b). Based on the results, I have proposed the [GADV]-protenoid microsphere hypothesis on the origin of cell structure. Furthermore, I have confirmed that the [GADV]-protenoid microsphere satisfies the indispensable conditions for the origin of cell structure, random process and evolution to modern cell structure as shown in Fig. 4.10.

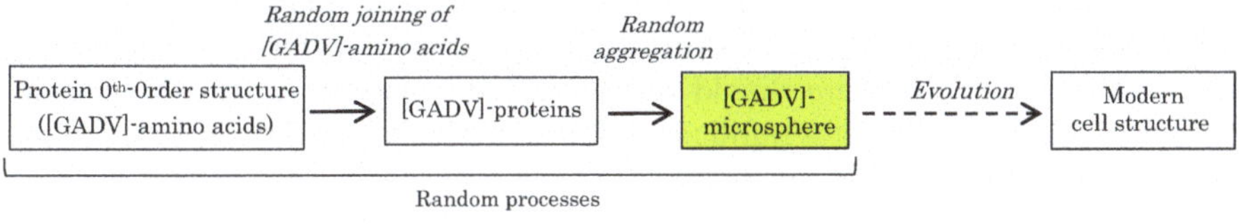

Fig. 4.10 [GADV]-microsphere was produced by random aggregation of [GADV]-proteins or actually [GADV]-peptides, which were synthesized by random joining of [GADV]-amino acids. In addition, [GADV]-microsphere could evolve to modern cell through the first genuine cell, which equipped with the six members constituting the fundamental life system.

5 Conclusion

Modern cell or organism is continuously divided and proliferates as supported by the fundamental life system composed of six members of gene, tRNA (genetic code), metabolism, cell structure and protein. Note that the six members are described in order of the flow of gene expression from gene to protein, because the five members are controlled by genes in modern organisms. On the other hand, a protocell or the most primitive cell structure can be regarded as a simple constituent of [GADV]-proteins, viewed from GADV hypothesis. My rough picture of protocell or cell is as follows.

1. Introduction of small compounds as H_2O and small organic compounds as glyoxylate, glyceraldehyde and pyruvate into the cell structure.
2. Synthesis of polymers as [GADV]-proteins and RNA with the small organic compounds introduced.
3. Swelling of the cell structure induced by increase of internal osmotic pressure upon synthesis of the peptides.
4. Division of the cell structure at a limiting point of growth of cell structure.
5. Evolution to cell division and proliferation under the fundamental life system.

Other features specific to the primitive protocell are further enumerated as follows.

1. The first protocell, which could be easily formed even on the primitive Earth, must be a spherical structure, because any specific growing point would not be required to construct the spherical protocell.
2. Protocell should be constructed easier, as the number of kinds of constituent molecules is smaller. Therefore, the [GADV]-microsphere, which was constructed with essentially one kind of [GADV]-peptides, should be the best for the first protocell.
3. Protocell could actively grow, if various compounds accumulated on the primitive Earth were incorporated into the inside.
4. The protocell, which was composed of [GADV]-peptides, would be the best, in the meaning that the protocell had also catalytic activities necessary to protein synthesis.

The expectations described above, of course, must be confirmed by experiments, because it cannot be always concluded that a thing with optimal properties for the protocell was actually the first cell structure. The essence of the primitive cell structure or [GADV]-microsphere should be maintained in modern cells or organisms through evolution for about 4 billion years from the [GADV]-microspheres to the modern organisms with fantastic features.

References

Baba T, Takeuchi F, Kuroda M et al (2002) Genome and virulence determinants of high virulence community-acquired MRSA. Lancet 359(9320):1819–1827

Deamer DW, Pashley RM (1989) Amphiphilic components of the Marchison carbonaceous chondrite: surface properties and membrane formation. Orig Life Evol Bioshp 19:21–38

Di Giulio M (2008) An extension of the coevolution theory of the origin of the genetic code. Biol Direct 3:37

Fox SW, Dose K (1977) Molecular evolution and the origin of life. Marcel Dekker Inc., New York

Fox SW, Jungck JR, Nakashima T (1974) From proteinoid microsphere to contemporary cell: formation of internucleotide and peptide bonds by proteinoid particles. Orig Life 5:227–237

Gilbert W (1986) The RNA world. Nature 319:618

Gorlero M, Wieczorek R, Adamala K et al (2009) Ser-his catalyses the formation of peptides and PNAs. FEBS Lett 583:153–156

Hua Z (2018) On the origin of life: a possible way from Fox's microspheres into primitive life. SOJ Biochem 4:1–7

Ikehara K (2009) Pseudo-replication of [GADV]-proteins and origin of life. Int J Mol Sci 10:1525–1537

Oba T, Fukushima J, Maruyama M et al (2005) Catalytic activities of [GADV]-peptides: formation and establishment of [GADV]-protein world for the emergence of life. Orig Life Evol Biosph 35:447–460

Oparin AI (1957) The origin of life on earth. Academic, New York

Prosser CL (1970) In Moore J (ed) Ideas in evolution and behavior, Natural History Press, Garden City, p 357

Shuwannachot Y, Rode BM (1998) Catalysis of dialanine formation by glycine in the salt-induced peptide formation reaction. Orig Life Evol Biosph 28:79–90

Wong JT (1975) A co-evolution theory of the genetic code. Proc Natl Acad Sci U S A 72:1909–1912

Chapter 5
The Origin of Metabolism

Abstract Modern metabolism is a network of chemical reactions, which are catalyzed by enzymes encoded by the respective genes. Several ideas for explaining the origin of metabolism have been proposed. For example, in some hypotheses, it is assumed that the proto-metabolism should be a chemical reaction system, in which the reactions were carried out with minerals, such as pyrites, clays and rocks on hydrothermal vents in a deep sea. In such cases, the activities on minerals must be transferred at one time to protein enzymes. However, it would be impossible to transfer the catalytic activities on minerals with three-dimensional structure to protein enzymes also with three-dimensional structure.

On the other hand, the five conditions must be satisfied to establish the most primitive metabolism as follows. (1) The first metabolism must be constructed through random process. (2) Chemical reactions, which were carried out in the initial metabolism, must be catalyzed by proteins produced through random process in the absence of gene, although those might be immature and incomplete. (3) Organic compounds, which were easily synthesized with prebiotic means and accumulated at large amounts on the primitive Earth, should be used as reactants in the initial metabolism. (4) In addition, organic compounds produced in the initial metabolism must be useful for the emergence of the first life. (5) Catalytic reactions proceeding at reaction steps as less as possible would be favorable for the initial metabolism.

Based on the analyses of modern metabolic reactions under the five conditions, three organic compounds, glyoxylate (Go), glyceraldehyde (Ga) and pyruvate (Pyr) (GoGaPyr) were selected out as reactants of immature [GADV]-proteins, which were produced by random joining of [GADV]-amino acids. The reasons are because the three GoGaPyr compounds composed of two or three carbon atoms could be easily synthesized with prebiotic means and should accumulate at large amounts on the primitive Earth, and also because it is supposed that the two organic compounds, glyoxylate and pyruvate, lead to syntheses of four [GADV]-amino acids (Gly [G], Ala [A], Asp [D] and Val [V]) and glyceraldehyde is a precursor molecule for synthesis of ribose 5-phosphate, which is necessary to produce nucleotides and RNA. The origin of metabolism proposed in this chapter is compatible with the idea that life emerged according to not the "gene/replicator-first" theory, but the "protein/metabolism-first" theory on the origin of life.

K. Ikehara, *Towards Revealing the Origin of Life*,
https://doi.org/10.1007/978-3-030-71087-3_5

Keywords Origin of metabolism · Surface metabolism theory · Clay world theory · Hydrothermal vent theory · [GADV]-amino acids · Protein 0th-order structure · GoGaPyr hypothesis

1 Introduction: Towards Solving the Riddle of the Origin of Metabolism

1.1 Necessary Matters for Exploration of the Origin of Metabolism

1. It is necessary to well understood what features modern metabolism has, in order to make clear the origin of metabolism, a member of the fundamental life system.
2. It is indispensable to understand the features of modern metabolism or answering to the following question, from what kind of primitive metabolism the contemporary metabolism originated.
3. Furthermore, an evolutionary process from the primitive metabolism, which could be formed by random process, to modern metabolism, must be explained reasonably.
4. If it is assumed that the contemporary metabolism originated from primitive one with features different from the contemporary metabolism, as a matter of course, the primitive metabolism must be transferred to the system having features similar to modern metabolism and further the system must evolve modern one, and the evolutionary process must be reasonably explained, either directly or indirectly (Figs. 5.1 and 5.2).

1.2 General Features of Modern Metabolism

About the Field of Catalytic Reactions

1. The modern metabolism is driven by proteinaceous catalysts or enzymes.

About the Metabolic Network

1. Modern metabolic system, which is driven by protein catalysts or enzymes, forms one network (Fig. 5.1).
2. Glycolysis, TCA cycle and pentosephosphate cycle are playing a central role in the modern metabolism.
3. Enzymes generally catalyze organic compounds, which have accumulated at a large quantity in a cell. Therefore, organic compounds, which were newly produced through catalytic reactions in previously developed metabolic system and accumulated in the cell, are generally used in new metabolic reactions, as expected by coevolution theory (Wong 1975; Wong et al. 2016; di Giulio 2008).

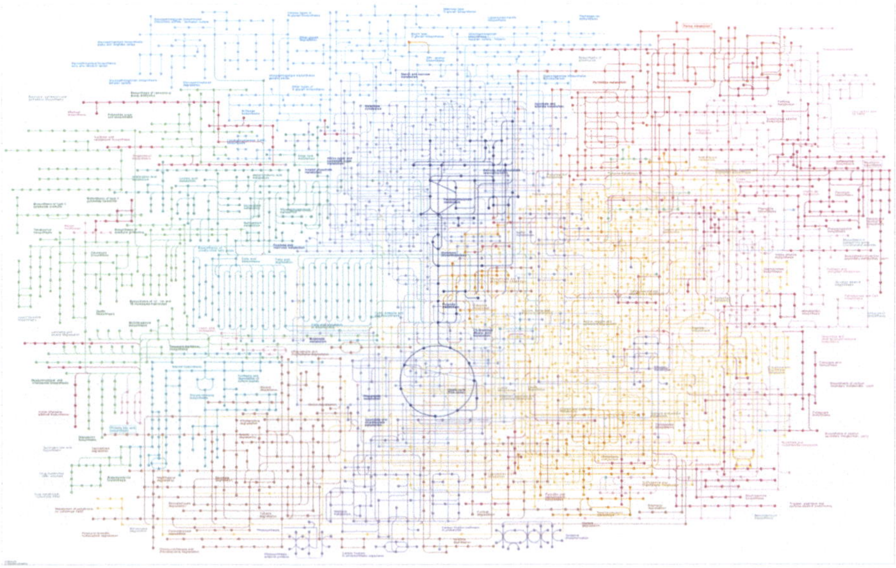

Fig. 5.1 Modern metabolic map. Glycolysis, TCA cycle and pentosephosphate cycle are highlighted with blue lines. The map was obtained from KEGG metabolic pathways (map01100). How was the first metabolic pathway created through random process on the primitive Earth? The answer for solving the riddle of the origin of metabolism is described in this chapter

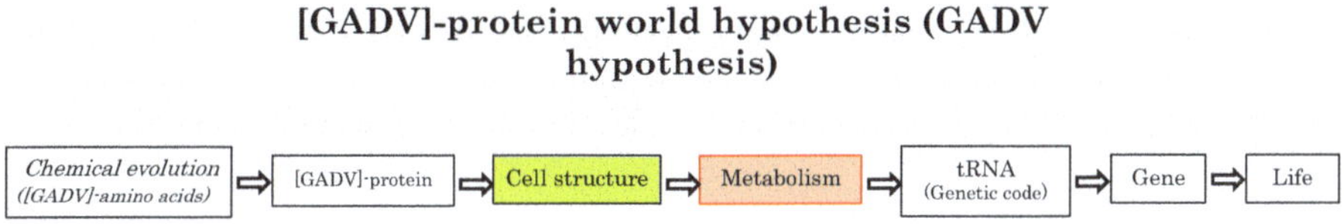

Fig. 5.2 In this chapter, the origin of metabolism is discussed (red box). The most primitive metabolic pathways would be formed in [GADV]-protein world or in [GADV]-protocell structure ([GADV]-protenoid microsphere) (yellow box)

1.3 Basic Features of Modern Metabolism

1. Metabolism starts from organic compounds produced by carbon dioxide assimilation of CO_2 and nitrogen fixation of N_2.
2. Organic compounds, into which CO_2 are incorporated, are chiefly used for syntheses of phosphorylated sugars, as glyceraldehyde 3-phosphate and ribose 5-phosphate in modern metabolism.
3. N_2 and NH_3 are first incorporated into organic compounds as aspartic acid, Asp.
4. Syntheses of amino acids, ribose and nucleotides are main metabolic reactions in modern metabolism.

5. Main products in the modern metabolic system, amino acids, ribose and nucleotides, are used for substrates of syntheses of protein, tRNA and gene composing the fundamental life system (Fig. 5.2).

1.4 Significance of Establishment of the Primitive Metabolic System for the Emergence of Life

1. Many researchers including supporters of RNA world hypothesis on the origin of life would admit that complex compounds like as nucleotides were not produced with prebiotic means at large amounts on the primitive Earth. Nevertheless, they might state that it is important to create the first gene with the small amounts of nucleotides for the emergence of the first life.
2. Then, assume that RNA could be produced with the small amounts of nucleotides. However, the nucleotides should be exhausted before the emergence of life, because sufficient amounts of nucleotides could not be supplied soon from the circumstances upon the consumption of nucleotides, because it is difficult to synthesize and accumulate nucleotides at large amounts with prebiotic means. Therefore, even if the first life emerged, proliferation of the life could not persist for a long time.
3. In that sense, it is reasonable to consider that life emerged under a primitive metabolic pathway, which was driven by immature [GADV]-proteins.

Two sides should be considered for exploration of the origin of metabolism. The first one is what catalysts were used as the reaction field of the first metabolism. The second one is from what catalytic reactions the first metabolism originated.

2 The Ideas on the Origin of Metabolism Advocated by Other Researchers

First, it is discussed from the first side on the origin of metabolism.

2.A. The Origin of Metabolism-I: What Catalysts Were Used in the First Metabolism?

The origin of metabolism, of course, has been studied by many researchers thus far. Consequently, several ideas, in which metabolism originated from catalytic reactions on surfaces of inorganic materials, such as pyrites (Wächtershäuser 1988), clays (Hartman 1975; Cairns-Smith 1977) and hydrothermal vents (Baross and Hoffman 1985; Russell and Hall 1997), have been proposed by other researchers based on the studies on what kinds of organic compounds could be produced by the reactions (Kitadai et al. 2019).

For example, it has been reported that oxalate and glyoxylate etc. could be synthesized from CO_2 and H_2O under a light emission by ion-rich clays, and that amino acids also could be produced through reverse TCA cycle in thioester world (Hartman and Smith 2019). Of course, various organic compounds produced with such prebiotic means should be utilized to synthesize amino acids and nucleotides, which are indispensable to the emergence of the first life. However, the studies have been too strongly focused on elucidation about what kind of organic compounds could be produced with the inorganic catalysts to explore the origin of metabolism. I discuss the ideas on the origin of metabolism presented by other researchers one by one in the following sections.

2.1 Surface Metabolism Theory

There exists many papers discussing the possible roles of minerals in the prebiotic stages of the chemical evolution (Bernal 1951; Cairns-Smith 1977; Wächtershäuser 1992). Among the original proposals, minerals have been considered in: (a) process that would discriminate molecular chirality; (b) condensation reactions of biomolecular precursors; (c) prebiotic catalysis; (d) biochemical templates; and (e) autocatalytic metabolism. It is also described that physicochemical changes of mineral surfaces – caused by environments simulating Archean aqueous scenarios – should be taken into account in the proposals of mineral prebiotic roles.

Strengths of Surface Metabolism Theory Undoubtedly, it is quite important for the studies on the origin of life to confirm where and what kinds organic compounds were produced on the primitive Earth. The surface metabolism theory, which was founded on the viewpoint, played an important role in the study on the origins of metabolism and life. The reason is because various organic compounds could be produced from simple inorganic compounds as CO_2, H_2O and N_2 through the catalytic reactions on minerals, such as pyrite (Wächtershäuser 1988). The organic compounds synthesized on the minerals should be used for syntheses of amino acids and proteins, which are indispensable to the emergence of life.

Weaknesses of Surface Metabolism Theory However, the catalytic center on the minerals never be transferred to proteinaceous catalysts or enzymes, meaning that the modern metabolic reactions did originate from the catalytic reactions on not minerals but protein.

2.2 Clay World Theory

The clay world hypothesis has been proposed by Hartman (1975) and Cairns-Smith (1977). In the hypothesis, the origin of metabolism is discussed based on replication of clays, and organic syntheses through reverse TCA cycle under a light emission.

The hypothesis has been presented from a standpoint of the bottom-up approach exploring what affairs happened on the primitive Earth.

Strengths of Clay World Theory In the paper of Hartman and Smith (2019), they have described that the self-replicating clays would photochemically fix CO_2 into organic acids and gradually evolve into the sulfide-rich region acquiring N_2 fixation in the process. A thioester world had come into being. The entry of phosphate into the evolving catalytic network expanded the metabolic network. The phosphate world had come into being. Thus, in the clay word theory, syntheses of various organic compounds have been confirmed by catalytic reactions on clays.

Weaknesses of Clay World Theory However, the catalytic activity on the clays cannot be also transferred to the surface of protein enzyme and, furthermore, the establishment process of the first life system composed of the first gene, genetic code and protein could not be clarified founded on the clay world theory. Therefore, the origin of metabolism could not be solved by the hypothesis.

2.3 Hydrothermal Vent Theory

Many researchers consider nowadays that hydrothermal vents may be plausible fields for the emergence of life, because hydrothermal vents can supply inorganic compounds, like as H_2 and CO_2, which are necessary to synthesize organic compounds, as CH_4, CH_3COOH and so on, in addition to supplies of energy and proton (Zhang et al. 2017; Preiner et al. 2019).

Strengths of Hydrothermal Vent Theory In fact, Zhang et al. (2017) recently described that in a novel hydrothermal system in Archean subseafloor model, highly alkaline and high temperature hydrothermal fluids were generated in basalt-hosted hydrothermal vents, where H_2 and CO_2 could be abundantly provided. These extreme conditions could have played an irreplaceable role in the early evolution of life. Furthermore, Zhang et al. (2017) have reported in their recent paper, that glycine could be synthesized from ethanolamine in simulated Archean alkaline hydrothermal vents.

Weaknesses of Hydrothermal Vent Theory However, even if methane, acetic acid, glycine and so on could be synthesized in the hydrothermal vents, and even if organic syntheses at the hydrothermal vents were similar to chemical reactions carried out in autotrophic microorganisms, it could not be concluded that modern metabolism originated from those catalytic reactions in/on the hydrothermal vents, because catalytic activities on hydrothermal vents cannot be also transferred to proteinaceous catalysts or enzymes, either directly or indirectly. This indicates that the modern metabolism did not originate from the hydrothermal vents.

2.4 Three Hypotheses Proposed by Other Researchers and the Origin of Metabolism

As described above, several ideas about the origin of metabolism, such as hypotheses based on catalytic reactions on pyrites, clays and hydrothermal vents, have been proposed. However, all of them have been studied from the viewpoint, on what kinds of organic compounds could be produced by catalytic reactions on surfaces of inorganic minerals. On the other hand, all metabolic reactions in modern organisms are driven by protein catalysts or enzymes. Therefore, the catalytic reactions on minerals must be transferred to those on the enzyme surfaces one day, if the hypotheses proposed thus far are correct. However, it would be impossible in principle to transfer the catalytic activities on minerals with a three-dimensional structure to the surfaces of three-dimensional proteins (Fig. 5.3).

As described above, many researchers have considered the reaction fields of the most primitive metabolism, as pyrites, clays and hydrothermal vents on the primitive Earth. However, I would like to insist that even the first primitive metabolic system must be established in the field surrounded by a [GADV]-protenoid membrane or in [GADV]-microsphere to protect from dissipation of organic compounds produced with the metabolic reactions, which were catalyzed by [GADV]-proteins.

The reasons, why [GADV]-proteins must be used in the primitive metabolic system in the [GADV]-microsphere, are as follows.

(1) [GADV]-amino acids could be used for synthesis of [GADV]-proteins for a long time, because [GADV]-amino acids accumulated at large amounts on the primitive Earth.
(2) It would be easy to supply [GADV]-amino acids through syntheses of [GADV]-amino acids from simple organic compounds, glyoxylate and pyruvate, which were easily produced with prebiotic means.
(3) Owing to those, various catalytic reactions could proceed through binding of [GADV]-proteins with the organic compounds, which were produced with prebiotic means and accumulated at large amounts on the primitive Earth, to produce [GADV]-amino acids, ribose, nucleotides, RNA and so on.

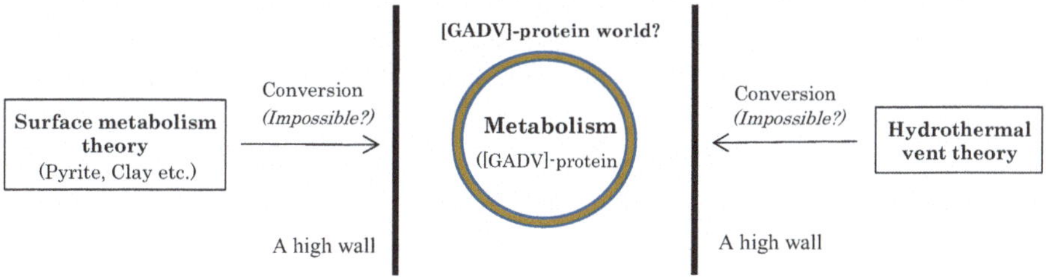

Fig. 5.3 The origin of metabolism have been investigated, about what kind of organic molecules could be synthesized from what kind of inorganic compounds, thus far. However, the riddle of the origin of metabolism could not be solved, if the following process could not be made clear, how catalytic functions on minerals such as pyrites, clays and hydrothermal vents could be transferred to protein catalysts (enzymes). However, it would be impossible to convert catalytic activities on minerals to enzymes in principle

Of course, supply of the organic compounds produced with the non-protein catalysts was indispensable especially in the early phase toward the emergence of life. Hence, it is quite important to study about what kind of organic compounds could be produced by such catalytic reactions on surfaces of minerals and accumulated on the primitive Earth, because it would be impossible to solve the riddles of the origin of life without the organic compounds produced by the prebiotic means. On the other hand, it is well known that many organic compounds could be obtained by the so called "Miller's experiments", which were carried out by electronic discharge into the supposed primitive atmosphere. Therefore, it would be better to consider that the catalytic reactions on minerals are roughly equivalent to the Miller-type experiments, which should largely contribute to the emergence of life through supply of simple organic compounds.

Nevertheless, the studies on the origin of metabolism must rely on the non-proteinaceous catalytic reactions, although it would be unavoidable, because researchers providing such ideas have not recognize the immature [GADV]-proteins with some catalytic activities (Fig. 5.4). However, the origin of metabolism could not be solved forever, as long as the researchers hold fast to the idea, which is put on the basis of non-proteinaceous catalysts. This also indicates that the concept of protein 0th-order structure is indispensable to solve the origin of metabolism.

In addition, there is one more serious problem in the three hypotheses. That is, contemporary metabolism is carried out in a cell surrounded by cell membrane. Therefore, catalytic reactions on the surface of inorganic materials must be transferred into the inside of cell structure one day. However, it would be also impossible to transfer the catalytic system driven on surface of minerals into the cell structure. This means that the initial metabolism must be carried out in a cell structure carrying protein catalysts from the beginning (Fig. 5.4).

Many readers of this book might like to ask me "Then, how could protein be produced before creation of the first gene?" and "Could the proteins, which were produced in the absence of any genetic function, really catalyze many metabolic reactions?". I consider the answer to the questions as follows. The problems could be solved, if the readers can understand my idea, assuming that many chemical reactions could be catalyzed by immature [GADV]-proteins produced under the protein 0th-order structure, as already described in Chap. 3: Sect. 4.2 in detail.

This is one of the reasons, why many researchers have considered the origin of metabolism based on the catalytic reactions other than protein enzymes thus far. However, inversely stating this, that has made it difficult to solve the riddle of the origin of metabolism, because of the prejudice that any protein must be always

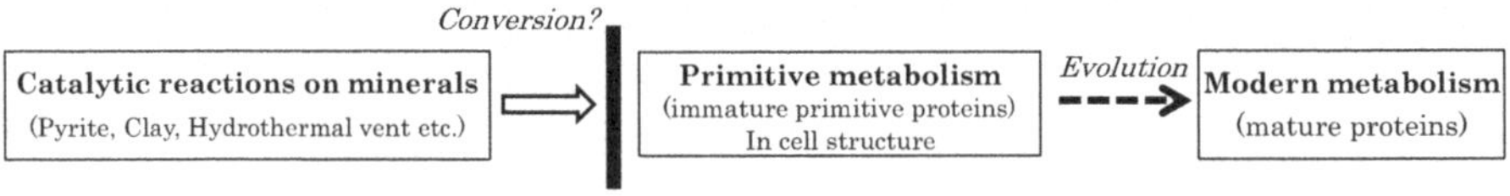

Fig. 5.4 The ideas, that catalytic reactions should be carried out on minerals, as pyrites, clays, hydrothermal vents and so on, before catalytic reactions driven by protein enzymes were invented, have proposed by several researchers to explain the origin of metabolism. However, it would be obvious that the origin of metabolism cannot be explained by the catalytic reactions on minerals, because of discontinuity or a high wall (bold line) that catalytic reactions on minerals cannot be transferred to those with primitive proteins. Bold broken arrow indicates multi-evolutionary steps

produced under the genetic system. The prejudice has also made it difficult to understand the way how the proteins constituting the initial metabolism were produced before creation of genes or in the absence of genes (see also Chap. 3: Sect. 4.2). Then, my idea on the origin of metabolism is explained in the Sect. 3.

2.B. The Origin of Metabolism-II: From What Chemical Reactions Metabolism Originated?

Second one is the origin of metabolic reactions about what catalytic reactions were carried out in the primitive metabolic system.

It seems to me that the problem, from what chemical reactions the modern metabolism originated, has not been seriously discussed among other researchers thus far. The reason would be probably because other researchers were interested in too strongly what kind of reaction were proceeded at where on the primitive Earth to care about much from what kind of reactions the metabolism originated.

3 My Idea on the Origin of Metabolism

3.1 For What Purpose Was the First Metabolism Created?

For what purpose was the first metabolic system created? Was the metabolic system created to achieve a predetermined purpose? The first metabolic system would be naturally formed as a result that, a field or cell structure, effective catalysts (immature [GADV]-proteins) and organic compounds, which were necessary to proceed effectively chemical reactions, simply existed on the primitive Earth, but not for achievement of a purpose. The origin and evolutionary process of metabolism are discussed with such problems in mind in the following sections.

3.A. The Origin of Metabolism-I: What Catalysts Were Used in the First Metabolism?

It is considered by many researchers that modern metabolism originated from chemical reactions carried out by inorganic catalysts, as pyrites, clays, hydrothermal vents and so on. On the contrary, the metabolic reactions in extant organisms are driven by proteinaceous catalysts. Therefore, it is mainly discussed how the modern metabolism originated from chemical reactions with what kinds of catalysts and organic compounds accumulated on the primitive Earth in the following sections.

3.2 The Initial Metabolism Must Be Formed Before the Emergence of Life

Modern metabolism is a network of catalytic reactions, with which organic compounds necessary to live are synthesized, while unnecessary organic compounds are degraded. Therefore, elucidation of the origin of metabolism is quite important to solve the riddle of the origin of life, because the first life also would emerge with organic compounds, which were necessary for the first life to emerge and could be produced through the initial metabolism.

Many readers may wonder whether [GADV]-amino acids and nucleotides could be really synthesized with immature [GADV]-proteins produced in the absence of any genetic system. Then, consider what events happened to the first life, which emerged in the absence of metabolic system. (1) Complex organic compounds, nucleotides, would accumulate only at a quite small amount on the primitive Earth, even if nucleotides could be produced with prebiotic means (Fig. 5.5). (2) In addition, it would be difficult to supply nucleotides into the first life because nucleotides would not be synthesized at large amounts with prebiotic means. (3) Therefore, nucleotides would be exhausted soon after RNA synthesis for production of tRNAs and genes started before or after the emergence of life. (4) Descendants from the first life would cease to proliferate soon because of the deprivation of nucleotides necessary to live in the absence of metabolic system (Fig. 5.5).

On the contrary, organic compounds composed of a small number of carbon atoms were preferentially synthesized with prebiotic means and accumulated at a large amount on the primitive Earth (Chap. 3: Fig. 3.8) (Miller and Orgel 1974). Therefore, the concentration of the three compounds, glyoxylate (Carbon number = 2), glyceraldehyde (C = 3) and pyruvate (C = 3) (GoGaPyr) in the circumstances would be almost unchanged, even if the three compounds were used for syntheses of [GADV]-amino acids, ribose and nucleotides in [GADV]-microspheres, because syntheses of the three simple organic compounds should proceed outside of the microspheres efficiently and continuously. It would be also obvious that, taking the ratio of external volume to internal volume of the microsphere into consideration, the concentration of the three GoGaPyr compounds outside of the microspheres should be essentially invariable. Therefore, it would be easy to supply the three GoGaPyr compounds into the [GADV]-microsphere for a sufficiently long time, even if a large quantity of the compounds were consumed in the microsphere.

Accumulation of organic compounds with prebiotic means naturally should increase at an arithmetic rate. On the contrary, proliferation of organism would

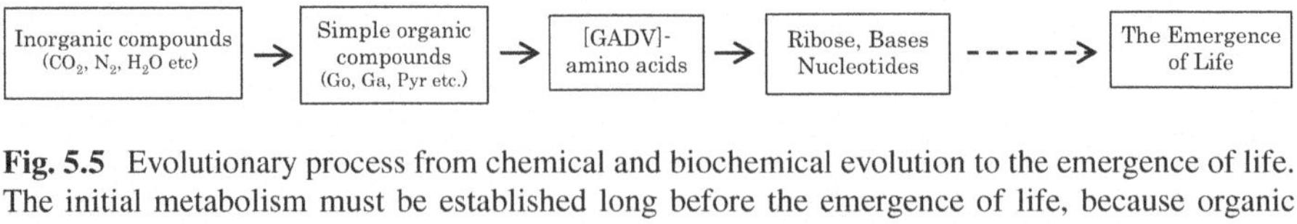

Fig. 5.5 Evolutionary process from chemical and biochemical evolution to the emergence of life. The initial metabolism must be established long before the emergence of life, because organic compounds, especially as nucleotides, which are necessary to synthesize tRNAs and RNA genes, could not been always synthesized at sufficient amounts by prebiotic means and should be exhausted at one time after all before the emergence of life. Therefore, complex organic compounds as nucleotides must be supplied through catalytic reactions in the initial metabolism, which were carried out with immature [GADV]-proteins in protocell

increase at an exponential rate. Therefore, organic compounds in organisms would be consumed at an exponential rate upon propagation of the organisms (Fig. 5.5). Of course, it is supposed that a proliferation rate of the [GADV]-microsphere should be quite low. Nevertheless, the exponential proliferation rate would reach at a quite high growth rate after several tens of generations. Microspheres growing at the high rate would consume nucleotides rapidly in parallel. Only one way for compensation of the organic compounds consumed in microspheres growing exponentially is to utilize a metabolic system in the microspheres, in which was driven by immature [GADV]-proteins, because the amounts of organic compounds synthesized in microspheres growing exponentially also increase at an exponential rate accompanied by the proliferation of microspheres as a whole.

However, there is another problem, that the metabolic system must continue to synthesize necessary organic compounds in the microspheres, even when they grew at a quite low rate. For that, microspheres must be stable to continue the syntheses of organic compounds in the microshpere, because the syntheses in the microspheres would eventually terminate, if unstable microshperes were degraded and organic synthesis in the microspheres stopped. For the problem, I preliminary confirmed by change of turbidity with UV irradiation that [GADV]-microspheres are enough stable for more than 1 month.

Many readers may still have doubts about whether a primitive metabolic system could be formed for syntheses of nucleotides and RNA in [GADV]-microspheres. However, I would like to lay stress on that every theory on the origin of metabolism should face with the same problems discussed up to here, because an exponential growth and the consumption of organic compounds necessary to live should begin just after the emergence of proto-organisms and life. Therefore, it would be obvious that a kind of metabolic system must be established before the emergence of the first life, although it is needless to say that many experiments must be done hereafter to confirm that the ideas presented above are correct. Then, discuss in the next section by what catalysts, which was produced before the emergence of life or in the absence of gene, the initial metabolism was driven.

3.3 The Role of [GADV]-Amino Acids or Protein 0th-Order Structure in the Origin of Metabolism

Even if the organic compounds accumulated on the surroundings at large amounts, many of them might be exhausted at one time after all before the emergence of life, if the initial metabolic system had not been established, as described above (Fig. 5.5). This means that an initial metabolic system, which was driven by immature [GADV]-proteins, must be established before the exhaustion of [GADV]-amino acids, and that the initial metabolism would originate from reactions for synthesis of [GADV]-amino acids using the immature [GADV]-proteins as catalysts, because the most important metabolites should be [GADV]-amino acids, which were

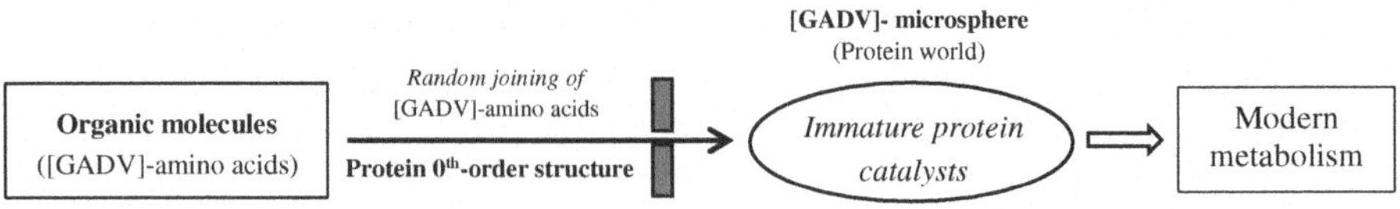

Fig. 5.6 Not mature but immature [GADV]-protein catalysts could be produced by random joining of [GADV]-amino acids, if synthesis of the immature proteins was carried out in the specific amino acid composition composed of [GADV]-amino acids or one of protein 0th-order structures. That is, the protein 0th-order structure makes a hole for entering into the protein world, in which the initial metabolism could be developed to synthesize [GADV]-amino acids and [GADV]-peptides

required to produce [GADV]-protein catalysts and for the protocells to proliferate continuously as driving force of [GADV]-peptide synthesis in the [GADV]-microsphere (Fig. 5.6).

On the other hand, the prejudice, that it would be impossible to synthesize protein catalysts without gene, made it difficult to get the way how to acquire a sufficient amount of nucleotides. In addition, metabolic reactions carried out even in the beginning of metabolism or in the most primitive metabolic system, must be catalyzed by proteins, because any transition pathway from non-proteinaceous catalysts on minerals to protein catalysts could not be discovered thus far (Figs. 5.3 and 5.4). It suggests that even the first metabolism should originate from a chemical reaction system, which was driven by not mature but immature proteins. The gist of my idea is as follows. First one is protein 0th-order structure or a specific amino acid composition, in which immature but water-soluble-globular proteins could be produced even by random joining of amino acids at a high probability (Chap. 3: Sect. 4.2). Protein 0th-order structure holds the first key for solving the enigma of the origin of metabolism. On the other hand, chemical reactions on minerals could supply [GADV]-amino acids and other organic compounds necessary to establish the initial metabolism, similarly to organic syntheses with other prebiotic means like as the so called "Miller's experiments". In other words, the establishment process of the initial metabolism or the origin of metabolism could be reasonably explained only by connecting protein catalysts produced under the protein 0th-order structure or [GADV]-amino acid composition with organic compounds synthesized by various prebiotic means (Fig. 5.6).

3.B. The Origin of Metabolism-II: From What Chemical Reactions Metabolism Originated?

In succession of the discussion of the first problem; "what catalysts were used in the initial metabolism?", the origin of metabolism-II; "from what chemical reactions did metabolism originated" is discussed in this section.

General Features of the Initial Metabolic Reactions

In spite of the significance of the initial metabolism, it is totally unknown at the present time point about what kind of chemical compounds were used as reactants in order to synthesize what kind of organic compounds. Therefore, one of main objectives of this chapter is to make clear these points in order. Then, the necessary conditions for establishment of the initial metabolism are first enumerated in

Table 5.1 Three conditions for integration of catalytic reactions into the initial metabolism. Of course, various organic compounds, which were unnecessary for the first life to emerge, should be also synthesized with immature [GADV]-proteins. However, microspheres producing unnecessary organic compounds at large amounts would be exterminated by selection from other microspheres

1. Use of organic compounds accumulated at a large amount on the primitive Earth.
2. Synthesis of organic compounds indispensable to the emergence of life.
3. Use of the organic compounds produced at less reaction steps.

Note that the reason, why the three conditions are necessary for formation of the initial metabolism, is because organic compounds used as reactants cannot be selected out and metabolic pathways cannot be designed in advance

Table 5.1, before exploration of the origin of metabolism-II. As described in Table 5.1, first of all, it would be important to make clear what catalytic reactions proceeded with what kind of organic compounds as reactants, and to confirm whether the products obtained through the catalytic reactions were actually required for activities of the first life. Furthermore, it is also important to examine the number of reaction steps from the starting material to the products, because it would be difficult to form a multi-step metabolic pathway before creation of genes. Such metabolic reactions satisfying the conditions were searched for from within the modern metabolic map as described below.

3.4 What Kinds of Catalytic Reactions Were Used in the Initial Metabolism

It is easily supposed that organic compounds, which accumulated at a larger amount on the primitive Earth than others, would be used in the initial metabolic reactions, because protein catalysts, which were produced in the absence of genetic function, cannot select out organic compounds as substrates with an intention.

3.4.1 What Organic Compounds Were Accumulated on the Primitive Earth?

Then, I explored first what organic compounds would accumulate on the primitive Earth at a larger amount than others, and were used in the initial metabolism, based on the results obtained by Miller and Orgel (1974). It is well known that organic compounds were synthesized at a larger amount in the so called "Miller's experiments", as the number of carbon atoms in the compound becomes smaller (Chap. 3: Fig. 3.8) (Ikehara under review). Cleaves et al have also reported that the results similar to those obtained by the Miller's experiments, which were carried out with a reducing atmosphere (CH_4, NH_3, H_2O, H_2), were obtained in the experiments of the electric discharge into a rather neutral atmosphere (CO_2, N_2, H_2O) (Cleaves et al. 2008).

It is well known that various kinds of organic compounds were prebiotically synthesized and accumulated on the primitive Earth, for example, by electric discharge into primitive atmosphere, by chemical reactions on minerals as pyrites, clays and hydrothermal vents (Sect. 2.A) and by introduction of organic compounds with meteorites, comets, interplanetary dust particles from space and so on (Chyba and Sagan 1992).

Therefore, the initial metabolic system would be composed of catalytic reactions using chemical compounds having a few carbon atoms (two or three) as reactants, because what organic compounds should be used in the system could not be selected previously (Table 5.1). In facts, the catalytic reactions using chemical compounds with a few (from two to six) carbon atoms as glyoxylate, pyruvate and sugars can be seen on the most fundamental and important modern metabolic pathways as glycolysis and tricarboxylic acid (TCA) cycle and their vicinity (Ikehara under review).

3.4.2 What Organic Compounds Were Required for the First Life to Emerge?

Not only organic compounds, which were indispensable to establishment of the fundamental life system composed of gene, genetic code and protein, but also organic compounds, which were unnecessary to the emergence of life, should be produced by catalytic reactions with primitive immature [GADV]-proteins. However, among all catalytic reactions catalyzed with the immature proteins, the reactions producing organic compounds required to progress to the emergence of life would be naturally integrated into the initial metabolism. Among them, amino acids and nucleotides, which were used as the respective reactants for syntheses of protein/peptide and RNA, should be the most indispensable for the first life to emerge (Table 5.2). This suggests that vestiges should remain even in the modern metabolic system, because all metabolites produced in the modern metabolic system are also indispensable for extant organisms to live.

Under the premises enumerated in Table 5.2, I would like to discuss further the origin of metabolism-II, how the first metabolic system was established using what kind of chemical compounds. The most important matter for the emergence of life is to establish the fundamental life system, which was composed of three main members, gene, tRNA (genetic code) and protein. Among the three matters, it could be stated that protein but not gene would play the central actor in the system, because

Table 5.2 Metabolic reactions producing organic compounds necessary for the first life to emerge

1. Synthesis of [GADV]-amino acids and [GADV]-proteins, actually aggregations of [GADV]-peptides, for obtaining catalytic activities (protein, cell structure, metabolism).
2. Synthesis of ribose and adenine, which are required to synthesize ATP for obtaining chemical energy to enlarge metabolic pathways, if necessary (energy).
3. Synthesis of nucleotides necessary to synthesize tRNA and RNA gene for obtaining the genetic function (tRNA (genetic code), gene).

proteins are exhibiting almost all biological functions and because gene and genetic code have been only supporting the protein synthesis through the life system, although some researchers working in the research field of the origin of life might consider it inversely as the gene-first.

Therefore, the first one, which is required for the first life to emerge, would be amino acids necessary to synthesize protein catalysts producing all organic compounds required for the first life to emerge (Table 5.2). Many researchers have also approved that [GADV]-amino acids could be easily produced by prebiotic means (van der Gulik et al. 2009; Higgs 2009). It has been also confirmed that water-soluble globular proteins could be formed by random joining of [GADV]-amino acids (Oba et al. 2005). However, even if a sufficient amount of [GADV]-amino acids could be produced with prebiotic means and accumulated on the primitive Earth, the amino acids might be exhausted before the emergence of life, if the amino acids could not be supplied through the initial metabolism (Fig. 5.5). Therefore, first of all, metabolic reactions producing [GADV]-amino acids must be integrated into the initial metabolism (Table 5.2). It is important to recognize again that even immature [GADV]-proteins, which were synthesized by random joining of [GADV]-amino acids under the one of protein 0th-order structure, could express various catalytic functions owing to the immature and pluripotent [GADV]-proteins with some flexibility, although the activities would be low but sufficiently high to lead to the emergence of life (Chap. 3: Sect. 4.2) (Oba et al. 2005).

Then, the chemical compounds with from one to three carbon atoms were selected out from the fundamental modern metabolic map (Nishizuka (Ed.), 1994) (Ikehara under review), as enumerated below.

Organic compounds with one carbon atom: formic acid, carbamate (carbamoyle phosphate)

Organic compounds with two carbon atoms: acetic acid (acetyl-CoA), glyoxylate (Gly)

Organic compounds with three carbon atoms: pyruvate (Ala) and organic compounds from glyceraldehyde to pyruvate in glycolysis

Note that organic compounds in parenthesis are derivatives from the precursor molecule described outside the parenthesis. On the other hand, both Gly and Ala out of four [GADV]-amino acids can be synthesized by addition of amino residue to the respective ketoacids, glyoxylate and pyruvate, at one reaction step (Nishizuka (Ed.), 1994). Therefore, the first and the second metabolic reactions, which were integrated into the initial metabolism, would be Gly synthesis with glyoxylate and Ala synthesis with pyruvate (Fig. 5.7). Asp and Val can be produced from glyoxylate at 3 steps and from pyruvate at 4 steps, respectively. Therefore, the synthetic pathways of [GADV]-amino acids would be integrated into the first phase of the initial metabolism (Fig. 5.7).

As can be seen in Fig. 5.7, Asp can be synthesized by amination of oxaloacetate, which is produced by dehydrogeation of malate synthesized from glyoxylate and acetic acid (acetyl-CoA) at one reaction step (Nishizuka (Ed.), 1994). Although Asp is also produced through incorporation of CO_2 into pyruvate in the modern metabolism, it would be difficult to proceed the reaction with immature [GADV]-proteins

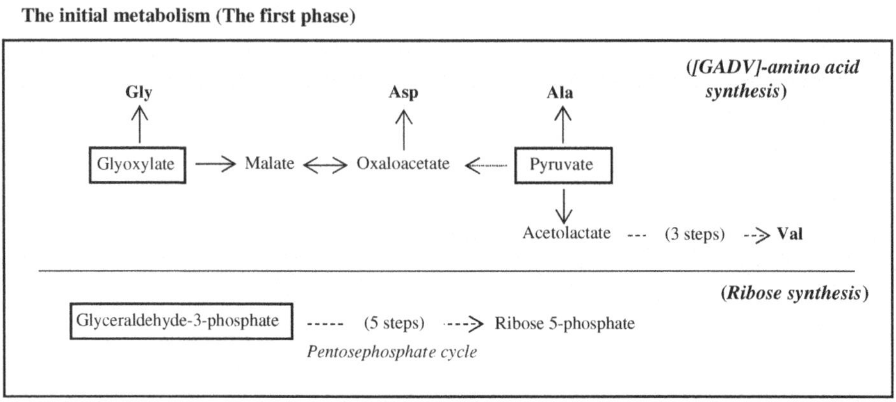

Fig. 5.7 The initial metabolism (the first phase) deduced based on the conditions (Tables 5.1 and 5.2) for the initial metabolism. (1) Gly is produced by amination of glyoxylate at one reaction step. (2) Ala is also produced by amination of pyruvate at one step. (3) Asp is synthesized from glyoxylate at three-step reaction through malate and oxaloacetate. (4) Val is synthesized from pyruvate at four reaction steps through acetolactate. Thus, synthetic pathways of [GADV]-amino acids were established to form the first phase of the initial metabolism. Ribose 5-phosphate synthesis from glyceraldehyde 3-phosphate is also included in the first phase of the initial metabolism. Solid arrows indicate that the reactions proceed at one step. Dotted arrow and broken arrows show the reaction step difficult to proceed and metabolic reactions proceeding at multi-steps, respectively

as aggregates of [GADV]-peptides because of the complex carboxylation. Therefore, it is supposed that Asp should be synthesized with oxaloacetate produced from glyoxylate through malate by three-step reaction (Fig. 5.7). Furthermore, Val is synthesized by amination of isovalerate, which can be produced at two-step reaction from acetolactate synthesized from pyruvate at one-step reaction (Nishizuka (Ed.) 1994). Therefore Val synthesis would become a limiting step of [GADV]-protein synthesis, because the four-step reaction is required to synthesize Val carrying five carbon atoms (Nishizuka (Ed.) 1994). These support the idea that the synthetic pathways of Asp and Val were integrated at a later stage than those of Gly and Ala into the initial metabolism (Fig. 5.7). Continuous accumulation of [GADV]-proteins through synthesis of Asp from glyoxylate and Val from pyruvate, which were incorporated from the circumstances around the [GADV]-microsphere, further accelerated the syntheses of more complex compounds, Asp and Val.

3.4.3 Nucleotide Synthetic Pathways Were Integrated at the Second Phase of the Initial Metabolism

Metabolic pathway necessary to synthesize RNA also must be formed, because ribose, nucleobases and especially nucleotides, which were required to RNA synthesis, were not accumulated with prebiotic means at large amounts on the primitive Earth and, therefore, should be exhausted before the emergence of life (Fig. 5.5).

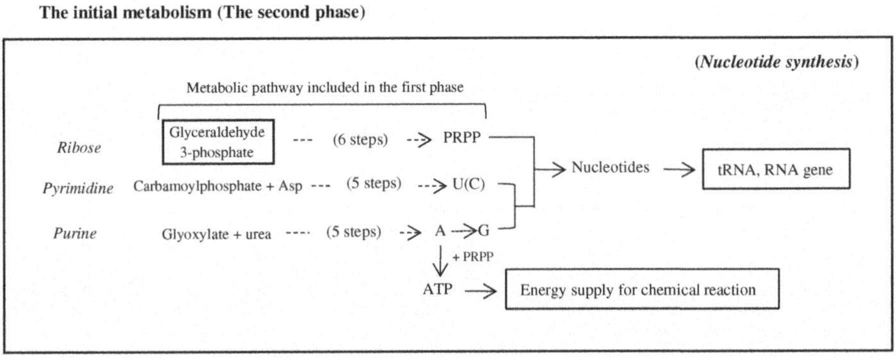

Fig. 5.8 The initial metabolism (the second phase), in which nucleotides and RNA were produced. The second phase is composed of three synthetic pathways; (1) pyrimidine base (uracil (U) and cytosine (C)) synthetic pathway at five reaction steps, which start from addition of carbamoly-phosphate to Asp. (2) It is supposed that adenine (A), which was produced from glyoxylate and urea through reverse salvage pathway, was used for ATP synthesis. Thus, RNA synthesis became possible after the second phase of the initial metabolism was established. ATP synthesized with adenine and PRPP was used to accelerate various chemical reactions. PRPP synthetic pathway from glyceraldehyde 3-phosphate (the first phase) is also included in this figure, because the pathway was used for nucleotide synthesis

Therefore, catalytic reactions synthesizing ribose with glyceraldehyde 3-phosphate and four bases should be incorporated into the second phase of the initial metabolism (Fig. 5.8). The reason is because glyceraldehyde 3-phosphate is an organic compound with only three carbon atoms and was one of starting organic compounds for synthesis of tRNA and RNA gene (Table 5.3). Successively, a large amount of ATP could be obtained by direct addition of adenine to ribose (actually, phosphoribosyl 1-pyrophosphate; PRPP), if ribose was produced through the first phase of the initial metabolism (Fig. 5.7). The number of reaction steps from glyceraldehyde 3-phosphate to ribose 5-phosphate is much less than the number of reaction steps from glucose and from pyruvate to ribose (Nishizuka (Ed.) 1994). This also supports the idea that glyceraldehyde 3-phosphate is the starting organic compound for synthesis of ribose 5-phosphate. Thus, four nucleotides could be synthesized with ribose 5-phosphate, which is necessary to synthesize nucleotides (Table 5.3 and Fig. 5.8).

(1) ATP alone, which was synthesized with adenine and phosphoribosyl pyrophosphate (PRPP), could supply chemical energy progressing catalytic reactions. However, even the synthesis of PRPP, an intermediate for nucleotide synthesis, requires the six-step reaction from glyceraldehyde 3-phosphate. This means that the metabolic pathway for formation of ATP would be incorporated at a later phase than synthetic pathways of [GADV]-amino acids, which are included in the first phase.
(2) Five reaction steps are necessary to synthesize pyrimidine nucleotide, UMP, which is also one of intermediates for syntheses of tRNA and RNA gene.

Table 5.3 Three organic compounds used as starting compounds in the initial metabolism

1. Glyoxylate for synthesis of Gly and Asp	leading to [GADV]-protein synthesis
2. Pyruvate for synthesis of Ala and Val (Asp)	
3. Glyceraldehyde 3-phosphate for synthesis of ribose 5-phosphate	
-------------------> leading to syntheses of ATP, tRNA and RNA gene	

(3) Asp, which is synthesized in the first phase of the initial metabolism, is used as starting compound for the synthesis of pyrimidine nucleotide, UMP, meaning that synthesis of UMP became possible after a sufficient amount of Asp could be accumulated in a [GADV]-microsphere.

3.5 Proposition of GoGaPyr Hypothesis on the Origin of Metabolism

Here, I would like to name the novel idea on the origin of metabolism as "GoGaPyr (three compound) hypothesis", assuming that the modern metabolism originated from the three simple organic compounds, glyoxylate, glyceraldehyde and pyruvate. If the hypothesis is correct, it would be expected for the following matters to occur.

1. A new metabolic pathway should be formed by a new reaction using an organic compound, which was produced in previously existed metabolic pathways, which were expanded from the three organic compounds, glyoxylate, pyruvate and glyceraldehyde, as the "coevolution theory *" expects (Wong 1975; Wong et al. 2016; di Giulio 2008). The growth of metabolic network makes it possible to produce new organic compounds stimulating the evolution of organisms on the Earth. For example, development of the primitive metabolic pathways, which originated from the three GoGaPyr compounds, could produce many organic compounds, such as [GADV]-amino acids, [GADV]-peptides, sugars containing ribose, nucleotides and RNA.

   * Coevolution theory: The theory has been proposed that evolutionary process of the correspondences between amino acids and codons in the genetic code were determined by the sequential appearance of new amino acids within the primor-

dial biochemical system. Therefore, it is assumed that the evolutionary pathway of the initial metabolism progressed according to the sequential appearance of organic compounds in [GADV]-microsphere in accordance with the coevolution theory. The coevolution theory, which can deduce the sequential appearance of metabolic reactions, regardless of evolutionary stages of the metabolic system, would be one of the important concepts in the life sciences.
2. Therefore, no metabolic pathway should be formed independently of the previously existed metabolic network. In other words, all metabolic pathways are integrated into one metabolic network, because all modern metabolic pathways have been developed using at least one of organic compounds, which were produced in previously existed metabolic pathways.

Hence, it could be also concluded that the hypothesis is correct, if the above properties expected from the GoGaPyr-three compound hypothesis are found, and if the formation process of the modern metabolism cannot be well explained, when any one of the three compounds is removed.

3.5.1 Validity of GoGaPyr Hypothesis on the Origin of Metabolism

I confirm validity of the GoGaPyr hypothesis from view points of irreversible or essentially irreversible reaction processes.

1. It is clear that the modern metabolic system did not originate from one organic compound, glyceraldehyde, because synthetic pathway of pyruvate from glyceraldehyde requires 5 reaction steps and both glyoxylate and pyruvate could be easily synthesized with prebiotic means and accumulated at large amounts on the primitive Earth (Nishizuka (Ed.) 1994).
2. Inversely, glyceraldehyde, which is indispensable to synthesize ribose, nucleotides and also RNA, could not be produced, if the metabolism originated from one compound, pyruvate or glyoxylate, because of the quite high energy barrier between the reaction step from pyruvate to phosphoenol pyruvate (Nishizuka (Ed.) 1994).
3. Finally, let us consider the interconversion between pyruvate and glyoxylate. Glyoxylate could not be produced from pyruvate, because the reaction step from malate to glyoxylate is irreversible. Pyruvate could not be also obtained from glyoxylate, because there is another essentially irreversible transformation step from oxaloacetate to pyruvate (Nishizuka (Ed.) 1994).

Judging from the above comprehensive discussion, it is reasonable to consider that the modern metabolism originated from catalytic reactions using immature [GADV]-proteins and the three GaGoPyr organic compounds, as starting compounds of the initial metabolism. After establishment of the initial metabolic system with immature [GADV]-proteins, the system could develop new metabolic pathways and the new pathways were incorporated into the system, when necessary.

3.5.2 Synthetic Reactions Preferentially Proceeded in the Initial Metabolism

It is well known that degradation products of many organic compounds return into TCA cycle in modern metabolism. However, degradation reactions of metabolites are excluded from the initial metabolic system, because degradation reaction of a compound would be always meaningless, if the compound is not excessively synthesized through the metabolic pathway and accumulated at a large amount in protocells and genuine cells. Therefore, degradation pathways should be formed at a later stage and added into a previously existed metabolic system, which was formed by a collection of synthetic pathways. Consequently, synthetic reactions were preferentially observed in the initial metabolic system, even if immature [GADV]-proteins had various degradation activities.

3.6 Immature Proteins, Which Were Produced in the Absence of Gene, Could Catalyze All Metabolic Reactions in the Initial Metabolism

As described in Chap. 3, it has been confirmed that immature but water-soluble globular [GADV]-proteins (actually, aggregates of [GADV]-peptides) could be produced by random joining of [GADV]-amino acids. I hypothesized so far that all the reactions in the first metabolic system could be catalyzed by the immature [GADV]-proteins before creation of the first gene or in the absence of any gene.

However, a quite difficult problem remains unexplained that immature [GADV]-proteins produced under the protein 0th-order structure ([GADV]-amino acids) could really catalyze all the reactions in the initial metabolism, as shown in Figs. 5.7 and 5.8. I answer the question as follows.

1. As I explained in detail in Chap. 3: Sect. 4.8.1, all mature proteins have been created from the respectively corresponding immature proteins, which were encoded by GC-$(GNC)_n$-NSF(a) carrying essentially random GNC codon sequence (Fig. 5.9a). Note here again that a mature protein never be created by random joining of amino acids at one stroke because of the extraordinary large sequence diversity of about 10^{130} (Chap. 3).
2. Therefore, all reactions, which are catalyzed by EntNew proteins or the first ancestor proteins of all protein families in all modern organisms, should be detected on surfaces of immature proteins, which were produced from codon sequences on antisense strands of GC-rich genes (GC-NSF(a)s).
3. Similarly, all catalytic activities used in the initial metabolism should be found on surfaces of immature [GADV]-proteins-1, which were produced by random joining of [GADV]-amino acids. I explain the reason in more detail with Fig. 5.9b. Then, consider formation process of the initial metabolic pathways, which should be driven by mature [GADV]-proteins (Fig. 5.9b). All metabolic

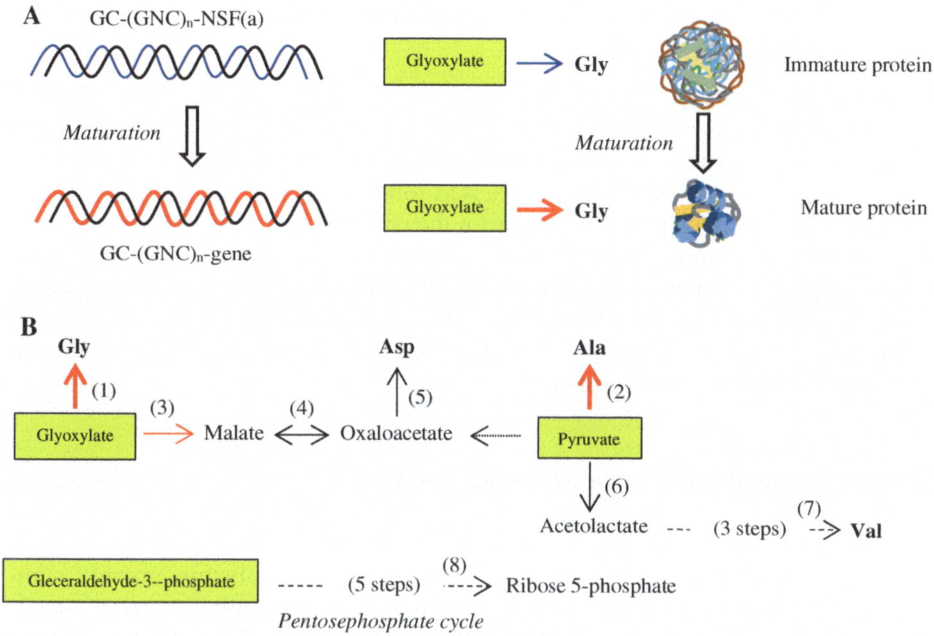

Fig. 5.9 (**a**) A mature protein should be created from an immature protein, which was produced from GC-(GNC)$_n$-NSF(a) (blue wavy line), on which essentially random GNC codon sequence is arranged. Synthetic reaction of Gly from glyoxylate is shown as one example. Note that mature protein never be created by random joining of [GADV]-amino acids at one stroke because of the extraordinary large sequence diversity of about 10^{60} (see also Chap. 3). (**b**) It is shown in the figure that two enzymes, Gly and Ala synthetases were already maturated from the respective immature [GADV]-proteins (bold red arrows) and maturation process of malate synthetase is in progress (thin red arrow). Other reaction pathways catalyzed with mature protein are not formed as yet (black thin arrows). All metabolic pathways shown in the figure should be catalyzed by immature [GADV]-proteins, because, otherwise, all the metabolic reactions catalyzed by mature proteins could not be formed, if any immature protein has not the corresponding catalytic activity

reactions for syntheses of [GADV]-amino acids and ribose in the initial metabolic pathways should be catalyzed by immature [GADV]-proteins, because all the reactions could be catalyzed by mature [GADV]-proteins, which were created from the corresponding immature [GADV]-proteins after establishment of the genetic system as shown in Chap. 3: Fig. 3.13. Inversely, this clearly indicates that immature [GADV]-proteins can express all the catalytic activities driven by the corresponding mature [GADV]-proteins.

3.7 Evolution of Metabolic Pathways from the Most Primitive Metabolic System

How could the amazing modern metabolic pathways composed of more than several tens of thousands of chemical reactions (see Fig. 5.1) be formed from the most primitive metabolic pathway? My idea is as follows.

Based on the GoGaPyr hypothesis on the origin of metabolism-II, it is considered that the modern metabolic system originated from chemical reactions using three organic compounds, GoGaPyr or glyoxylate, glyceraldehyde and pyruvate, which were synthesized with prebiotic means and accumulated in circumstances on the primitive Earth (Table 5.1). Then, how was modern metabolic pathway formed from the initial metabolic system?

3.7.1 Formation of Linear Metabolic Pathway

After the initial metabolic system was formed, the metabolic pathways would elongate randomly as incorporating new reactions catalyzing organic compounds, which were newly synthesized through the elongating metabolic pathways and accumulated in a [GADV]-microsphere or cell (Fig. 5.10). Catalytic reactions, which could produce organic compounds necessary for organisms to live, were retained in the evolving metabolic system. Thus, some linear

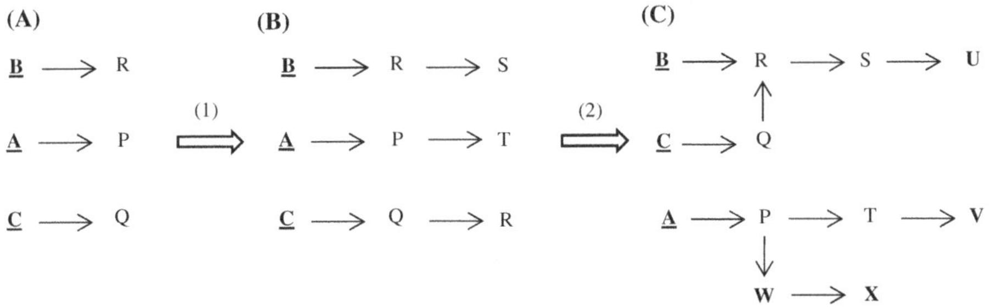

Fig. 5.10 Evolution of metabolic system: (**a**) In this example, it is shown that a metabolic system originated from chemical reactions catalyzing three organic compounds, **A**, **B** and **C**. (**b**) Thereafter, (1) the three metabolic pathways elongated accompanied by acquiring new enzymes to catalyze new organic compounds on the elongating metabolic pathways. (**c**) A branched pathway could be formed, (2) when a metabolite, R, which is the same intermediate of the other metabolic pathway, was accidentally produced on one independent pathway or when new metabolic reaction started from an intermediate on the previously existed metabolic pathway

metabolic pathways were formed. Usually, such a linear metabolic pathway having an objective compound at the one end would be formed, when the compound useful for organisms to live could be accidentally produced in the elongating reaction pathway.

3.7.2 Formation of Circular Metabolic Pathway

On the other hand, a circular metabolic pathway would be formed, when a common metabolite accidentally appeared at the two ends of a linear metabolic pathway elongating toward the both sides. This means that metabolism never originate from the cyclic metabolic pathway, as TCA cycle, because it is impossible to design such a cyclic metabolic pathway in advance. Therefore, it is supposed that TCA cycle started from probably malate synthesized from glyoxylate. And the TCA cycle could be completed, when a common compound, possibly 2-ketoglutarate, was produced at the two ends of the original linear metabolic pathway expanding from the starting molecule (malate) to both sides (see also Sect. 3.4).

3.7.3 Formation of Branched Metabolic Pathway

Branched metabolic pathways were formed in two different ways. (1) Two linear metabolic pathways elongating independently were connected through a common metabolite, when the metabolite, which was previously synthesized as an intermediate of one metabolic pathway, was produced at the end of the other elongating metabolic pathway (Fig. 5.10c). (2) Another formation process of branched pathway could start, when an intermediate, which had been previously produced on one linear or one circular metabolic pathway, was used as a substrate for formation of a new catalytic reaction. Therefore, a new branched pathway was formed by elongation of the new branch, which started from the intermediate (Fig. 5.10c).

Thus, the complex and amazing modern metabolic system was established through repeated elongation of linear metabolic pathways, formation of circular metabolic pathways and growth of branched pathways as accompanying selection of catalytic reactions necessary to live (Fig. 5.1).

Strengths of GoGaPyr Hypothesis It could be concluded according to the general necessary conditions (Sect. 1.1) for the origin of something that GoGaPyr hypothesis is valid as the origin of metabolism, because both formation process of initial metabolism through random processes and evolutionary process from the initial metabolism to modern metabolism can be explained as follows.

It is considered that the initial metabolism could be formed with three GoGaPyr organic compounds or glyoxylate, glyceraldehyde and pyruvate, which were accumulated by random processes on the primitive Earth and with immature

[GADV]-proteins, which were also produced by random joining of [GADV]-amino acids. It is also considered that metabolic pathways have evolved with catalytic reactions starting from the three GoGaPyr organic compounds and eventually the beautiful modern metabolic system has been completed (Fig. 5.1).

Weaknesses of GoGaPyr Hypothesis I have considered that the initial metabolism started from reactions, which were catalyzed by immature [GADV]-proteins using three GoGaPyr compounds as substrates. However, simultaneously I have also considered that there is a great weakness of the GoGaPyr hypothesis as that several matters about the origin of metabolism have not been confirmed with experiments as described below.

1. Could [GADV]-amino acids and [GADV]-petides be really synthesized by immature [GADV]-proteins in the initial metabolism in [GADV]-microsphere?
2. Could oligonucleotides and RNA be produced through the initial metabolism? and so on.

4 Discussion

There are two sides in the problem on the origin of metabolism. First one is "What catalysts were used in the first metabolism?", and second one is "From what chemical reactions did metabolism originate?".

Then, first discuss the problem on the origin of metabolism-I.

As described in Sect. 2.A, surface metabolism theory, clay world theory and hydrothermal vent theory have been proposed by other researchers to explain how modern metabolism has been established from what catalytic reactions. In all the three ideas, catalytic reactions on surface of minerals, as pyrites, clays and hydrothermal vents, are considered as the origin of metabolism. On the other hand, as well known, the modern metabolism is driven by proteinaceous catalysts or enzymes. Therefore, it would be impossible to consider catalytic reactions on the surface of minerals as original metabolic reactions, because the catalytic reaction on mineral with three-dimensional structure could not be transferred to the catalytic site of enzyme with tertiary structure (Sect. 2.4). Nevertheless, only catalytic reactions on the surface of minerals have been proposed as original metabolic reactions thus far. The reason would be probably because the origin of metabolism has been considered in the prejudice that protein never be produced in the absence of genetic function or gene. On the contrary, it would be obvious that the origin of metabolism never be solved under the situation (Fig. 5.11a). The key for overcoming the difficulty should be protein 0th-order structure or [GADV]-amino acids as described in Sect. 3.1 of this chapter.

Next, discuss the second problem on the origin of metabolism-II, "From what chemical reactions did metabolism originate?".

I have not heard that other researchers have discussed from what catalytic reactions, which were driven by proteins, the initial metabolism started. On the contrary,

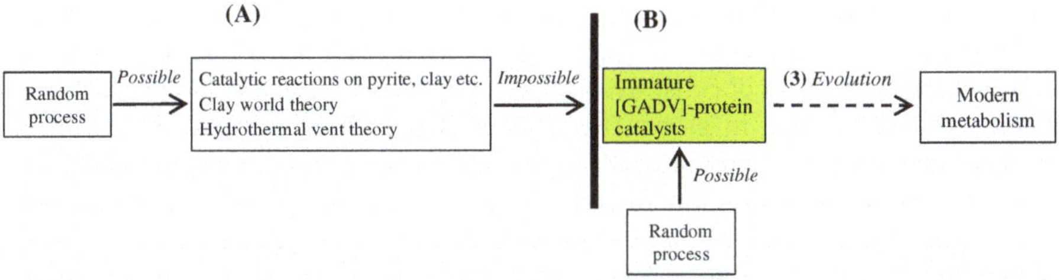

Fig. 5.11 (**a**) Catalytic reaction system on surfaces of pyrites, clays and hydrothermal vents could be formed through random processes. However, catalytic reactions on the surfaces could not be transferred to the surface of protein catalysts or enzymes. (**b**) In the GoGaPyr hypothesis on the origin of metabolism-II, it is assumed that catalytic reactions were carried out with immature [GADV]-protein catalysts, which were produced by random joining of [GADV]-amino acids, and with three organic compounds, glyoxylate, glyceraldehyde and pyruvate, which were randomly produced with prebiotic means and accumulated on the primitive Earth. Furthermore, it can be reasonably explained that the modern metabolism has been established as a result of evolution from the initial metabolic pathways using the three organic compounds and immature [GADV]-proteins

I have proposed a novel idea on the origin of metabolism-II as the GoGaPyr three compounds hypothesis in this chapter (Ikehara under review). Then, it is discussed here, whether or not the GoGaPyr three compounds hypothesis is valid as the idea explaining the origin of metabolism-II. As can be seen in Fig. 5.11, only the GoGaPyr three compounds hypothesis can explain reasonably that three GoGaPyr compounds could be produced through random processes and modern metabolic pathways have originated from initial metabolic reactions using the three compounds as starting points (see also Sect. 3.2).

The Reason Why the Modern Metabolism Forms One Network

Finally, consider the reason why the amazingly complex and beautiful modern metabolism (Fig. 5.1), which originated from only three chemical reactions independently catalyzing the three respective GoGaPyr organic compounds, glyoxylate, glyceraldehyde and pyruvate, could form one network. It can be interpreted as follows. First, metabolic pathways, which originated from the three catalytic reactions, formed one small metabolic network, because a common metabolite could be synthesized on two metabolic pathways elongating from three GoGaPyr compounds with simple chemical structure at a high probability. Thus, first two metabolic pathways were connected with each other (Fig. 5.10). When another metabolic pathway were similarly connected with a common metabolite, one metabolic network was formed. The main reason, why one network could be formed, would be because the variety of organic compounds, which are generated from the three simple GoGaPyr compounds with a small number of carbon atoms, is small and, therefore, a common metabolite could be produced on two independent metabolic pathways at a high probability. Thereafter, metabolic pathways always branched off from one of organic compounds, which were produced in the one metabolic network and accumulated in a cell at a high concentration, and the branches elongated. Thus, metabolic system gradually evolved during elongation of the first three metabolic pathways, formation of new branches, fusion of two branches to form a metabolic cycle and so on. Sometimes, unnecessary metabolic pathways might be formed. However, cell structures or cells, which elongated unnecessary metabolic pathways, should be disadvantage to grow. Therefore, cell structures or cells, which threw away the unnecessary metabolic pathways, were selected and continued to proliferate. Consequently, cell structure or cell, which retained one splendid metabolic network composed of only useful metabolic reactions, was selected and survived. Metabolic pathways have been developed as one network from the very beginning to the present.

5 Conclusion

Modern metabolic pathways have formed an enormously and intricately intertwined network composed of a large number of metabolic reactions (Fig. 5.1). Therefore, the metabolic system is an assembly of catalytic reactions, which are driven by enzymes and are growing continuously like a living thing, as elongating the previously existed metabolic branches, as generating new organic compounds in the metabolic system, and as making new metabolic branches and cycles when necessary.

The modern metabolic system is a result that many metabolic reactions were formed and grew dynamically from the initial metabolism, which was composed of three catalytic reactions using three GoGaPyr organic compounds as the respective substrates.

The following factors contributed to the establishment of the modern metabolic pathways.

1. Formation of two types of new genes/proteins: one is formation of a homologous gene/protein and the other is creation of an entirely new gene/protein.
2. Accumulation of new organic compounds: New organic compounds, which were produced by new metabolic reactions and accumulated in a cell, triggered to form further new metabolic reactions. Thus, the metabolic system forming one network has grown as a tree spreads its branches. Thus. the amazing modern metabolic system has been formed as the results.

References

Baross JA, Hoffman SE (1985) Submarine hydrothermal vents and associated gradient environments as sites for the origin and evolution of life. Orig Life Evol Biosph 15:327–345

Bernal JD (1951) The physical basis of life. Routledge & Paul, London

Cairns-Smith AG (1977) Takeover mechanisms and early biochemical evolution. Biosystems 9:105–109

Chyba CF, Sagan C (1992) Endogenous production, exogenous delivery and impact-shock synthesis of organic molecules: an inventory for the origins of life. Nature 355:125–132
Cleaves HJ, Chalmers JH, Lazcano A et al (2008) A reassessment of prebiotic organic synthesis in neutral planetary atmosphere. Orig Life Evol Biosph 38:105–115
Di Giulio M (2008) An extension of the coevolution theory of the origin of the genetic code. Biol Direct 3:37
Hartman H (1975) Speculations on the origin and evolution of metabolism. J Mol Evol 4:359–370
Hartman H, Smith TF (2019) Origin of the genetic code is found at the transition between a thioester world of peptides and the phosphoester world of polynucleotides. Life (Basel) 9(69)
Higgs PG (2009) A four-column theory for the origin of the genetic code: Tracing the evolutionary pathways that gave rise to an optimized code. Biol Direct 24:4–16
Ikehara K (under review) The origin of metabolism and GADV hypothesis on the origin of life. In: Smoukov SK, Seckbach J, Gordon R (eds) Conflicting models for the origin of life [Volume 1 in the series Astrobiology Perspectives on Life of the Universe, Eds. Richard Gordon & Joseph Seckbach]. Wiley-Scrivener, Beverly
Kitadai N, Nakamura R, Yamamoto M et al (2019) Metals likely promoted protometabolism in early ocean alkaline hydrothermal systems. Sci Adv 5:eaav7848
Miller SL, Orgel LE (1974) The origins of life on the earth. Prentice-Hall, Englewood Cliffs
Nishizuka Y (ed) (1994) Tokyo kagakudojin. In: Taisha-Mappu (The metabolic map) (in Japanese)
Oba T, Fukushima J, Maruyama M et al (2005) Catalytic activities of [GADV]-peptiders. Formation and establishment of [GADV]-protein world for the emergence of life. Orig Life Evol Biosph 35:447–460
Preiner M, Xavier JC, Vieira AN et al (2019) Catalysts, autocatalysis and the origin of metabolism. Interface Focus 9:20190072
Russell MJ, Hall AJ (1997) The emergence of life from iron monosulphide bubbles at a submarine hydrothermal redox and pH front. J Geol Soc Lond 154:377–402
Van der Gulik P, Massar S, Gilis D et al (2009) The first peptides: the evolutionary transition between prebiotic amino acids and early proteins. J Theor Biol 261:531–539
Wächtershäuser G (1988) Before enzymes and templates: theory of surface metabolism. Microbiol Rev 52:452–484. 0146-0749/88/040452-33$02.00/0
Wächtershäuser G (1992) Groundworks for an evolutionary biochemistry: the iron-sulfur world. Prog Biophys Mol Biol 58:85–201
Wong JT (1975) A co-evolution theory of the genetic code. Proc Natl Acad Sci U S A 72:1909–1912
Wong JTF, Ng SK, Mat WK et al (2016) Coevolution theory of the genetic code at age forty: pathway to translation and synthetic life. Life (Basel) 6(12)
Zhang X, Tian G, Gao J et al (2017) Prebiotic synthesis of glycine from ethanolamine in simulated Archean alkaline hydrothermal vents. Orig Life Evol Biosph 47:413–425

Chapter 6
The Origin of tRNA

Abstract Several ideas have been proposed by other researchers to explain the origin of tRNA or how modern L-form tRNA was formed, like as the minihelix theory, the double-hairpin theory and so on. It seems to me that those hypotheses have been too strongly influenced by the structure of modern tRNA and consequently the strong influence instead has made it difficult to solve the origin of tRNA. On the contrary, I got a novel hypothesis, anticodon stem-loop (AntiC-SL) hypothesis on the origin of tRNA through base sequence analysis of 5′ stem sequences of anticodon stem-loop of *Pseudomonas aeruginosa* tRNAs (Ikehara K, Orig Life Evol Biosph 49:61–75, 2019). The hypothesis suggests that modern tRNA originated from a small but stable hairpin-loop RNA composed of 17 nucleotides, which was produced through random process. According to the hypothesis, it can be reasonably assumed that modern tRNAs originated from one nonspecific AntiC-SL, which can carry one of [GADV]-amino acids selected randomly. Further, it can be hypothesized without any large discrepancy as that four specific [GADV]-AntiC-SL tRNAs were formed from one nonspecific AntiC-SL tRNA through four nonspecific AntiC-SL tRNAs. The evolutionary process from the one AntiC-SL tRNA to L-form modern tRNA could be also deduced without any large contradiction.

Keywords Origin of tRNA · Minihelix theory · Double-hairpin hypothesis · tRNA split gene hypothesis · Hairpin-loop RNA hypothesis · Anticodon stem-loop (AntiC-SL) tRNA hypothesis

> It is possible to trace the entire polynucleotide chain with only two minor regions of ambiguity. The polynucleotide chain has a secondary structure consistent with the cloverleaf conformation; however, its folding is different from that proposed in any model. The molecule is made of two double-stranded helical regions oriented at right angles to each other in the shape of an *L*. One end of the *L* has the CCA accepter; the anticodon loop is at the other end, and the dihydrouridine and TψC loops form the corner.

About 50 years has already passed since A. Rich's group determined L-form tRNA structure (Kim et al. 1972). However, it is totally unknown still now, from what structure the first tRNA originated and how the first tRNA was created through random process, only which occurred on the primitive Earth.

K. Ikehara, *Towards Revealing the Origin of Life*,
https://doi.org/10.1007/978-3-030-71087-3_6

1 Introduction: Towards Solving the Riddle of the Origin of tRNA

Exploring the origin of tRNA is to understand from what primitive tRNA the modern tRNAs originated and how the modern tRNA was formed from the primitive tRNA (Figs. 6.1 and 6.2).

1.1 Necessary Matters for Exploration of the Origin of tRNA

1. It is necessary to well understand what features modern tRNA has, in order to make clear the origin of tRNA composing the fundamental life system.
2. It is indispensable to understand the features of modern tRNA for exploration of the following question, from what primitive tRNA the contemporary tRNA originated.
3. Successively, evolutionary process from the primitive tRNA, which could be created by random reactions on the primitive Earth, to modern tRNA must be explained reasonably.

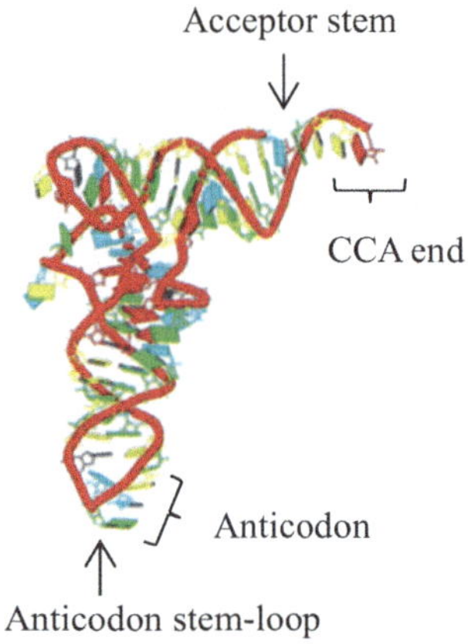

Fig. 6.1 Three-dimensional structure of yeast tRNA-Asp (PDBj: 1VTQ). Contemporary tRNA is skilfully folded into L-form, as can be seen in the right figure. Both anticodon stem-loop and accepter stem are completely exposed to the solvent, meaning that the two structures themselves are stable against hydrolysis by RNase

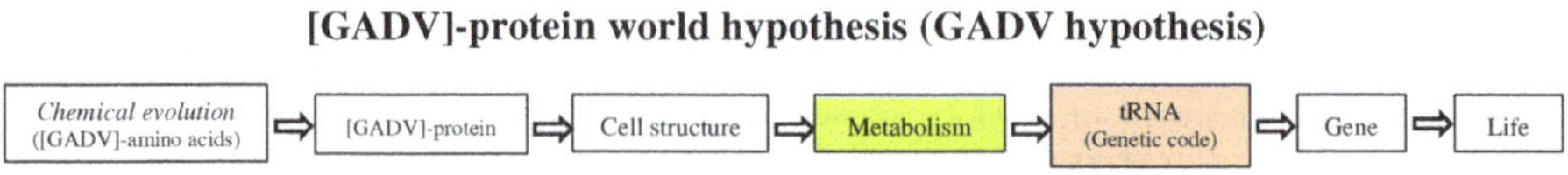

Fig. 6.2 In this chapter, the origin of tRNA is discussed (red box). Formation of the most primitive tRNA was supported by the initial metabolism, in which nucleotides and RNA could be synthesized by immature [GADV]-proteins (yellow box)

4. If it is supposed that the contemporary tRNA originated from primitive one with features different from the contemporary tRNA, as a matter of course, evolutionary process from the primitive tRNA to the primitive one with similar features to modern tRNA must be reasonably explained, either directly or indirectly.

1.2 General Structural Features of tRNA

1. tRNA is a rather small RNA composed of about 75 nucleotides.
2. The secondary structure of tRNA can be drawn like a clover with three leaves.
3. The secondary structure or the cloverleaf-form tRNA is further folded into a L-form tRNA stable against RNase.

1.3 General Functional Features of tRNA

1. The function of tRNA is to translate genetic information to amino acid sequence of a protein using anticodon and the amino acid binding site. In other words, tRNA assists protein synthesis using anticodon, which reads codon on mRNA, and amino acid bound with 3′-end of $^{3'}$-ACC-$^{5'}$of tRNA. Therefore, the distance between the anticodon and the amino acid bound with tRNA must be the same or at least similar among all tRNAs used to protein synthesis in one organism.

Discuss in this chapter the ways how the first primitive tRNA could be created and evolved to modern tRNA and how the correspondences between a codon/anticodon and an amino acid could be established.

2 The Ideas on the Origin of tRNA Advocated by Other Researchers

Then, consider whether or not the hypotheses on the origin of tRNA, which have been proposed by other researchers, are valid, as referring the features of modern tRNA described in Sects. 1.2 and 1.3.

As a matter of course, many researchers have tried to make clear the origin of tRNA up to the present time. Minihelix, double-hairpin and split tRNA gene hypotheses on the origin of tRNA are well summarized in the review by Fujishima and Kanai (2014). Therefore, I discuss whether or not the tRNAs assumed by the three hypotheses are suitable to the first primeval tRNA (Fig. 6.3).

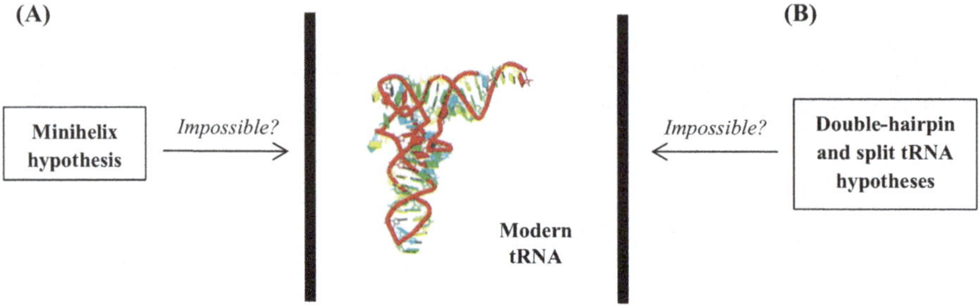

Fig. 6.3 (**a**) Did modern tRNA originate from minihelix composed of T-stem loop (T-SL) and accepter stem? However, it would be difficult to produce the minihelix at one stroke through random processes and also hard for the minihelix without anticodon to mediate between codon and amino acid. (**b**) Was the first tRNA constructed with two hairpins? However, it would be probably impossible to produced two hairpins at one stroke through random processes

2.1 Minihelix Theory

Minihelix theory advocates that the first tRNA originated from the minihelix (31-nt) composed of T-stem loop (T-SL) (17-nt containing 5 base pairs) and accepter stem (14-nt composed of 7 base pairs) (Tamura 2015). The facts that two helices of the T-SL and accepter stem are co-axial and that the minihelix can be aminoacylated by aminoacyl tRNA synthetase and the aminoacylated minihelix can participate in peptide bond formation in the ribosome have become the reasons supporting the minihelix theory (Schimmel and de Pouplana 1995; Tamura and Schimmel 2006; Tamura 2015). The hypothesis is also supported by the fact that the identity elements or "the second code" of tRNA are embedded within the accepter stem as the operational RNA code at few base pairs downstream from the end of the stem (Schimmel et al. 1993). Furthermore, Root-Bernstein et al. (2016) have proposed formation process for the conserved archaeal tRNA (75-nt), assuming that the 75-nt tRNA has evolved from ligation of three proto-tRNA minihelices (31-nt) and two-symmetrical 9-nt deletions within joined accepter stems.

Strengths of Minihelix Theory Modern tRNA as well as the most primitive tRNA should play a key role under the respective genetic systems (Chap. 2: Fig. 2.1a). The minihelix theory has been presented to clarify the origin of tRNA, which is one of the members of the fundamental life system. Several experimental results supporting the minihelix theory have been obtained by experiments, as described above.

Weaknesses of Minihelix Theory However, such facts do not always give evidence showing that modern tRNA originated from the minihelix, because it is natural that the minihelix, a part of modern tRNA, can be aminoacylated by aminoacyl tRNA synthetase and peptide bond can be formed with the aminoacylated minihelix in ribosome. Furthermore, it would be difficult to create the 31-nt minihelix through random process, which was composed of the large number of 12 base pairs, because the appearance probability by random joining of nucleotides is too small ($\sim 1/10^7$) to form the minihelix. In addition, it would be also impossible to read genetic information by the minihelix, because the minihelix has not anticodon which is indispensable to synthesize protein.

2.2 *Double-Hairpin and tRNA Split Gene Hypothesis*

Double-hairpin hypothesis and split tRNA gene hypothesis have been also proposed, assuming that tRNA was created by ligation of two duplicated fragments or half-sized hairpin like RNA structures (di Giulio 2006, 2008; Sugahara et al. 2009) and similarly by transcription of split genes encoding two half-sized modern tRNA (Tanaka and Kikuchi 2001).

Strengths of Double-Hairpin and tRNA Split Gene Hypotheses The double-hairpin theory has been proposed as focusing on the characteristic base sequences of two halves of modern L-form tRNA, which plays a key role in the fundamental life system, based on the fact that complementary sequences are observed between folded strands of the respective two halves of tRNA. Therefore, it has been assumed in the hypothesis that tRNA originated from two halves of tRNA

Weaknesses of Double-Hairpin and tRNA Split Gene Hypotheses However, it would be quite difficult to produce even the half-sized modern tRNA through random reactions, because the number of base pairs contained in the half-sized hairpin like RNA is estimated as about 10 base pairs and the probability of appearance roughly estimated is quite small as $\sim 1/10^6$ (Ikehara 2019). There is another problem about anticodon in the double-hairpin model that both two hairpin RNAs do not have anticodon and a functional anticodon appears at the last stage of L-form tRNA formation (Fig. 6.4).

In both cases, it would be quite difficult to produce even single hairpin structure of half-sized modern tRNA through random reactions, because of the smallness of appearance probability. Furthermore, the complementary relation between two halves of tRNA can be well explained by AntiC-SL tRNA hypothesis, as shown in Fig. 6.4c–e. In fact, a high complementarity in both D-loop hairpin and T-loop hairpin could be confirmed by integration of two bases of 5′ end of anticodon-loop and some bases of 3′ end of anticodon stem into spacer region between D-loop hairpin and T-loop hairpin using excess bases of variable region (Fig. 6.4d and E). One of the reasons would be because five base pairs are originally contained in the D-loop hairpin and T-loop hairpin. Another reason is because formation probability of base pairs in the remaining region, such as 5 base pairs, is high as $\sim 10^{-3}$. Therefore, formation of a large number of base pairs in the half-sized hairpin should be observed by analysis of tRNAs according to the double-hairpin model.

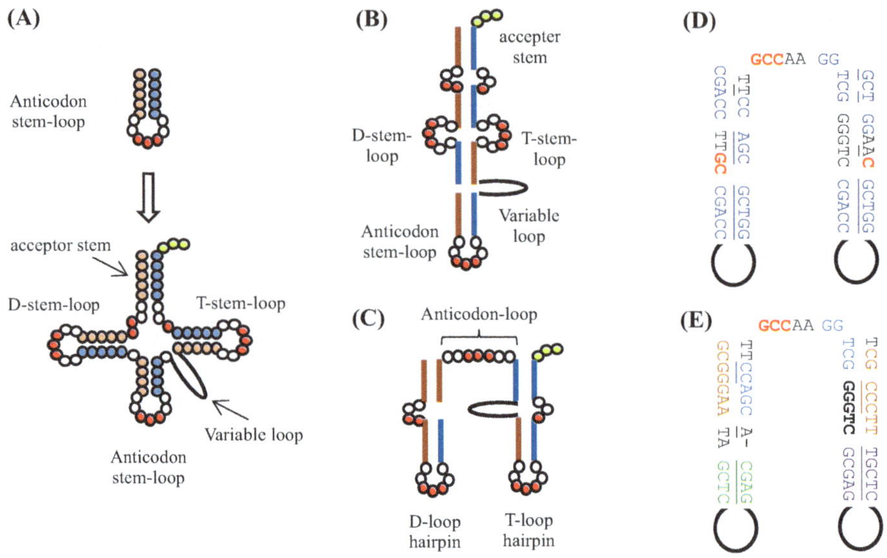

Fig. 6.4 (**a**) Assumed formation process of modern tRNA with four anticodon stem-loops (AntiC-SLs). AntiC-SL (top) and the clover leaf model of modern tRNA (bottom). It is possible to form modern tRNA by combination of three AntiC-SLs and one accepter stem derived from the original AntiC-SL (below). Complementary stem sequences are drawn with small brawn balls and blue balls in (**a**). (**b**) It can be seen from the figure that two halves of the tRNA, which is formed by combination of four AntiC-SLs, are roughly complementary each other. (**c**) A simple double-hairpin model of the tRNA, which is formed by combination of four AntiC-SLs. Complementary stem sequences are drawn with brawn lines and blue lines in (**b**) and (**c**). (**d** and **e**) Double-hairpin model viewed from AntiC-SL tRNA hypothesis. Taking the adjustment with some bases into consideration, the double-hairpin model can be explained pretty well based on the AntiC-SL tRNA hypothesis. Double-hairpin model was drawn using (**d**) four modern *Pseudomonas aeruginosa* Gly-AntiC-SLs and (**e**) one modern *P. aeruginosa* Gly-tRNA (tRNAdb (Universitateipzig)). Underline of base indicate base pair including G-T pair

2.3 A Small Hairpin-Loop RNA Hypothesis

Rodin et al. (1996) have proposed 11 base pair long hairpin-loop RNA as a candidate of the first tRNA based on reconstruction of the common consensus image of accepter stems from 1,268 tRNA sequences.

Strengths of a Small Hairpin-Loop RNA Hypothesis The hairpin-loop, which was proposed by Rodin et al. (1996), is similar to the AntiC-SL, which I have proposed as the most primitive tRNA in the AntiC-SL hypothesis (Ikehara 2019).

In their paper, they have also proposed a formation process from the 11 base palindrome to the modern tRNA-like cloverleaf structure. Their conclusion is also similar to the AntiC-SL hypothesis in the meaning of considering a small hairpin RNA as the first primeval tRNA. In addition, the probability of appearance, ~$1/10^{2.4}$, calculated from base pair number of the single hairpin constituting the double-stranded palindrome is considerably larger than that, ~$1/10^3$, of the AntiC-SL, which I have proposed in this chapter (Ikehara 2019).

Weaknesses of a Small Hairpin-Loop RNA Hypothesis However, there is a serious weak point in the small hairpin-loop RNA hypothesis, which Rodin et al. (1996) have considered, as described below. Although they have addressed that a single-stranded triplet in the loop of the small RNA became anticodon of tRNA, the loop composed of three bases is too small to arrange the three anticodon bases to make base pairs being opposite three codon bases in the loop of another hairpin-loop or of mRNA. Furthermore, "the second code" is placed in stem sequence of the hairpin-loop proposed by Rodin et al. (1996). However, the reason would be simply explained as that the stem was created by insertion of one primeval AntiC-SL into anticodon of another AntiC-SL (Ikehara 2019).

3 My Idea on the Origin of tRNA

3.1 For What Purpose Was the First tRNA Created?

Was tRNA created for translating a genetic information or for synthesizing a protein? If tRNA was created for translating a genetic information, the genetic information must exist before the first tRNA was created. If the assumption is correct, RNA, which was produced by random joining of nucleotides, must meet the genetic information for synthesis of a mature protein in the enormously vast nucleotide sequence space of about 10^{180}. That would be realistically impossible. Then, was the first tRNA created for the purpose of protein synthesis? If tRNAs were created for synthesizing a mature protein, an amino acid sequence, which is required to synthesize the mature protein, must be arranged by the tRNAs. This would be also realistically impossible for the tRNAs to arrange one specific amino acid sequence in the enormously vast amino acid sequence space of about 10^{130}. Therefore, tRNAs should be created simply in order to synthesize an immature [GADV]-protein more effectively than the synthesis by direct joining of [GADV]-amino acids. These mean that the creation order of three core members or protein, tRNA (genetic code) and gene (genetic information) had been previously determined in principle. In other words, there is nothing else besides that the three core members were created in order of protein, tRNA (genetic code) and gene (genetic information). Then, it is discussed what the first tRNA was and how the first tRNA was created.

3.2 Exploring the Origin of tRNA

As well known, tRNA is, generally, composed of three stem-loop (SL) structures (D-SL, anticodon (AntiC)-SL and T-SL), in addition to one accepter stem without loop structure (Fig. 6.5). The secondary structure of tRNA can be drawn as a clover with three leaves composed of the three SL structures and is, further, folded into tertiary L-form structure (Fig. 6.1).

When sequence similarity among 5′ stem sequences of *P. aeruginosa* tRNAs was analyzed to search for vestiges of the origin and evolution of tRNA (Fig. 6.5), it was fortunately found that all the 5′ AntiC-stem sequences of the tRNAs translating G-start codons are mutually and intimately related (Fig. 6.6a).

The results obtained implies that those 5′ AntiC-stem sequences of AntiC-SLs were formed by succeeding to a previously existed ancestor sequence. So, it can be interpreted that the results shown in Fig. 6.6b represent the origin and evolutionary process of AntiC-SLs or primeval tRNAs for translating G-start codons, because nine, actually ten, 5′ AntiC-stem sequences of *P. aeruginosa* tRNA for G-start codon were mutually related as one linear relation map, in which Glu (TTC) is branched off from Asp (GTC) (Fig. 6.6b). The mutual relation map contained connections of 5′ AntiC stem sequences for four C-start codons with the corresponding sequences for G-start codons, suggesting that those for C-start codons originated from the corresponding ones for G-start codons. The intimate mutual relationship also suggests that the vestiges of primeval tRNAs certainly remain among 5′ AntiC-stem sequences or AntiC-SLs of modern *P. aeruginosa* tRNA. Four, actually five, 5′ AntiC-stem sequences of tRNAs for GNC codon stand in a line as (Gly (GCC)-Val (GAC)-2Asp (GTC)-Ala (GGC)). Four 5′ AntiC-stem sequences of *P. aeruginosa* tRNAs for translating C-start codon (Leu (GAG1), His (GTG), Pro (GGG) and Pro (TGG)) are mutually related with one of the 5′ AntiC-SL tRNAs for G-start codon, respectively (Fig. 6.6b). In the figure, it is tentatively assumed that AntiC-SL tRNAs originated from the most ancestor Val-AntiC-SL tRNA.

However, I, thereafter, noticed that there exist several questions in the origin and evolutionary pathways of tRNAs, especially about the most primeval tRNA. Then,

	5' Accepter-stem		5' D-stem	D-loop	3' D-stem		5' AntiC-stem	AntiC-loop	3' AntiC-stem
5'-end	GCGGGAA	TA	GCTC	AGTT-GGT--A	GAGC	A	CGACC	TT<u>GCC</u>AA	GGTCG

GG---(---)---GTC	GCGAG	TTCGAGT	CTCGT	TTCCCGC	T	CCA	3'-end
Variable loop	5' T-stem	T-loop	3' T-stem	3' Accepter-stem		CCA end	

Fig. 6.5 DNA base sequence of modern *P. aeruginosa* PAO1 Gly-tRNA gene, which is composed of three stem loops and one accepter stem, is shown. Color is painted in the box according to the tRNA database (tRNAdb (Universitat Leipzig)) to easily discriminate among respective stem sequences. Brown, light green, blue, pink colors show accepter-, D-, anticodon (AntiC)-, T-stem sequences, respectively. Sequence in light blue box and underlined red letters indicate 3′-CCA terminal sequence and anticodon of the tRNA, respectively

(A)

5' Anticodon stem		
Gly	GCC	CGACC
Val	GAC	CCACC
2Asp	GTC	CCGGC
Ala	GGC	CTTGC
Ala	TGC	CCTGC
Val	TAC	TCTGC
Gly	TCC	TCAGC
Gly	CCC	TGAGC
Glu	TTC	CCGCC

(B)

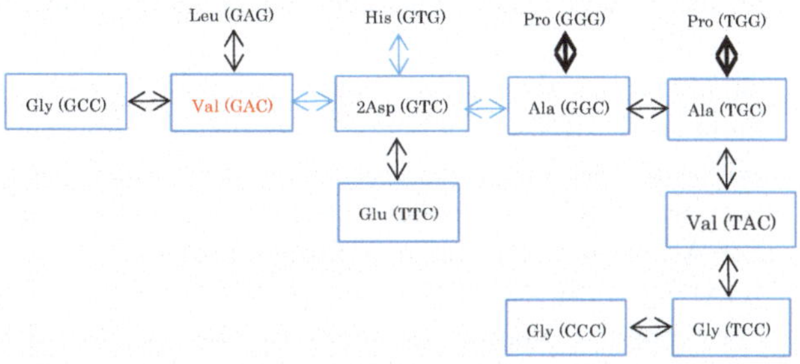

Fig. 6.6 (**a**) Amino acid, anticodon and 5′ AntiC-stem sequence are written in 1st, 2nd and 3rd columns from the left, respectively. 2Asp means that two 5′ AntiC-stem sequences of two Asp (GTC)s are the same. When one base is different between two anticodons or two 5′ AntiC-stem sequences, the base is shown with the same bold color letter. 5′ AntiC-stem sequences are intimately related to the sequence in one line below by one base difference, except Asp (GTC), which is weakly related with Val (GAC) and Ala (GGC) by two base differences, as enclosed with red rectangles. The square bracket shows that Asp (GTC) and Glu (TTC) are mutually related by one base difference. (**b**) Mutual relation map among nine, actually ten, 5′ AntiC-stem sequences of *P. aeruginosa* tRNA for translating G-start codon. Two pairs of tRNAs with sequence similarity were connected with black bold, black thin and blue thin double headed arrows, when the number of conserved bases in the stem sequences is five, four and three out of the five bases, respectively. It is tentatively assumed in the Fig. 6.6b that AntiC-SL tRNAs originated from the possible first ancestor Val-AntiC-SL tRNA written with red bold letters in the map

I propose a more reliable origin of the first AntiC-SL tRNA and an evolutionary pathway from it to the five primeval AntiC-SLs for [GADV]-amino acids and Glu for answering the questions in the next section.

3.3 Exploring the First Primeval AntiC-SL tRNA

As described in the above section, it was confirmed that the first primeval tRNAs were four AntiC-SLs carrying anticodon GNC. However, several questions still remain unanswered on the origin of tRNA especially on the first AntiC-SL tRNA, which it is difficult to explain reasonably. Therefore, it is discussed what the first AntiC-SL was.

Question 1. What aminoacyl (aa)-AntiC-SL was the first AntiC-SL tRNA?

Question 2. Furthermore, anything should generally originate from one something. However, one AntiC-SL must be useless to synthesize [GADV]-protein and the useless AntiC-SL should be abandoned at one time.

Question 3. The results shown in Fig. 6.6b indicate that a group of AntiC-SLs for GNC codon had already reached at a sufficiently high evolutionary level, because all the 5′ AntiC-stem sequences are mutually connected with other AntiC-stem sequences by one or two base differences, meaning that one 5′ stem sequence should be already discriminated from others by a protein or a primitive [GADV]-aminoacyl tRNA synthetase (pri-ARS). However, it would be naturally impossible to form a gene encoding a primitive ARS, which can recognize the specific AntiC-stem sequence before creation of at least four specific AntiC-SL tRNAs. On the other hand, a plural number of specific tRNAs, for example four [GADV]-AntiC-SL tRNAs for [GADV]-protein synthesis, always require the corresponding number of genes for the specific ARSs. This means that there exists a "chicken and egg relationship" between tRNA and ARS, which would be quite difficult to explain formation process of the relationship, because gene encoding ARS should require multiple tRNAs for the gene expression. On the contrary, tRNAs carrying specifically one of [GADV]-amino acids should be totally useless in the absence of mature ARSs, which are encoded by the respective genes.

Thus, the origin and evolutionary pathway of AntiC-SL tRNA meet with at least three questions as described above. The only one way for answering the three questions is to assume that the four [GADV]-AntiC-SLs originated from one nonspecific co-ancestor AntiC-SL, which can bind with one of [GADV]-amino acids randomly or nonspecifically. Then, I searched for the co-ancestor AntiC-SL according to the following procedure (Fig. 6.7).

1. First, 5′ AntiC-stem sequence of a hypothetical co-ancestor of two progeny AntiC-SLs, Val (GAC) and Asp (GTC) or of Asp (GTC) and Ala (GGC), was presumed. On the ground of that 5′ AntiC-stem sequences of Val (GAC) and Asp (GTC) or of Asp (GTC) and Ala (GGC) are mutually related by two base differences, (1) I searched for a base sequence of co-ancestor connecting with the two progenies by one base difference. Two hypothetical 5′ AntiC-stem sequences were obtained, because the two progenies are mutually related by two base differences (Fig. 6.7). (2) Next, base differences between the 5′ AntiC-stem assumed and those among the respective four progenies, Gly (GCC), Val (GAC), Asp (GTC) and Ala (GGC), were counted. (3) I selected one 5′ AntiC-stem sequence connecting with minimum number of base differences from the co-ancestor to the four progenies. From the results, CCAGC could be selected as 5′ AntiC-stem sequence of the coancestor (Coa)-[GADV]-AntiC-SL tRNA (Fig. 6.7).
2. Next, ancestor or precursor (pre)-5′-AntiC-stem sequences of Gly (GCC)- and Ala (GGC)-AntiC-SLs were deduced, both of which are connected with the CCAGC stem sequence of the Coa-[GADV]-AntiC-SL tRNA and with those of the progenies, Gly (GCC)- and Ala (GGC)-AntiC-SLs, by one base difference. From the results, two pathways connecting the Coa-[GADV]-AntiC-SL with all the six progenies by one base difference were obtained as can seen in Fig. 6.7. Successively, one nonspecific AntiC-SL, which carries one of four GNC anticodons and can randomly bind with one of [GADV]-amino acids at 3′ CCA

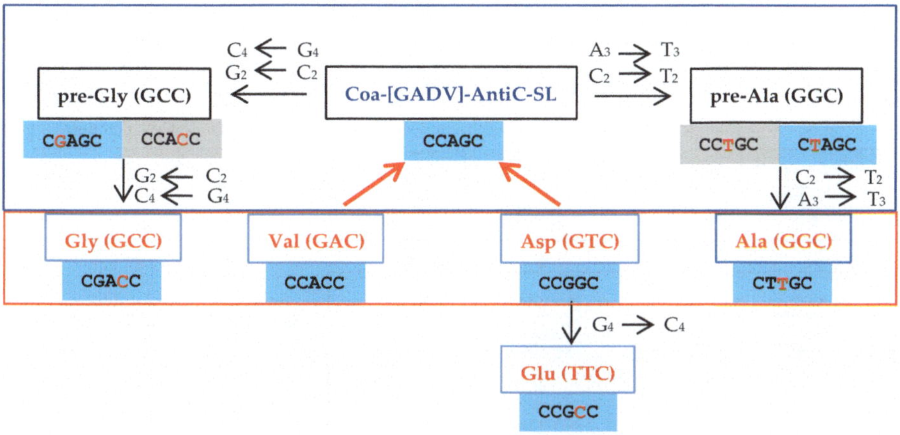

Fig. 6.7 Deduced origin of AntiC-SL tRNAs and evolutionary pathway from the first nonspecific to five specific AntiC-SLs for translating GNC and GAR codon into five amino acids ([GADV]-amino acids and Glu [E]). 5′AntiC-stem sequences of the first nonspecific Coa-[GADV]-AntiC-SL was deduced with two 5′ AntiC-stem sequences of Val (GAC) and Asp (GTC) tRNAs as shown with bold red arrows. Two evolutionary pathways from the first AntiC-SL tRNA to specific Gly (GCC)-, or Ala (GGC)-AntiC-SL tRNA enclosed in blue rectangle can be theoretically considered as shown by two 5′ AntiC-stem sequences described in blue and gray color boxes of pre-Gly (GCC) and pre-Ala (GGC). Subscript indicates base position, at which base of AntiC-stem was replaced. Base of 5′AntiC-stem sequence replaced from the ancestor sequence one step before is shown with red letter in blue or gray color boxes. Anticodon and 5′AntiC-stem sequences are written as DNA sequence, because genes encoding the respective tRNAs were used for the analysis (tRNAdb (Universitat Leipzig))

end, was selected out. Note that both aminoacyl-AMP and aminoacyl-$^{3'}$ACC$^{5'}$ were non-specifically used as activated forms of [GADV]-amino acids for [GADV]-protein synthesis.

Furthermore, four kinds of nonspecific AntiC-SLs could be derived from the first nonspecific AntiC-SL (Fig. 6.8). If it was achieved, single-stranded (ss)-(GNC)$_n$ gene could be produced by random joining of GNC anticodons in AntiC-loops of the four nonspecific AntiC-SLs (Fig. 6.9). Then, double-stranded (ds)-RNA carrying (GNC)$_n$ random sequence could be created through complementary strand synthesis of the first ss-RNA. Consequently, immature [GADV]-protein with random [GADV]-amino acid sequence, which was produced by expression of ds-RNA having (GNC)$_n$ random sequence, could evolve to mature protein with ordered [GADV]-amino acid sequence, owing to a memorizing ability of base substitutions on the ds-(GNC)$_n$ RNA. In this way, a mature ARS protein, with which one AntiC-stem sequence can be discriminated from other AntiC-stem sequences, could be produced by using the ds-(GNC)$_n$ RNA for the first time. Furthermore, the creation of four mature [GADV]-primeval (pri)-ARSs could stimulate establishment of four AntiC-SL tRNAs with specific 5′ AntiC-stem sequence through co-evolution

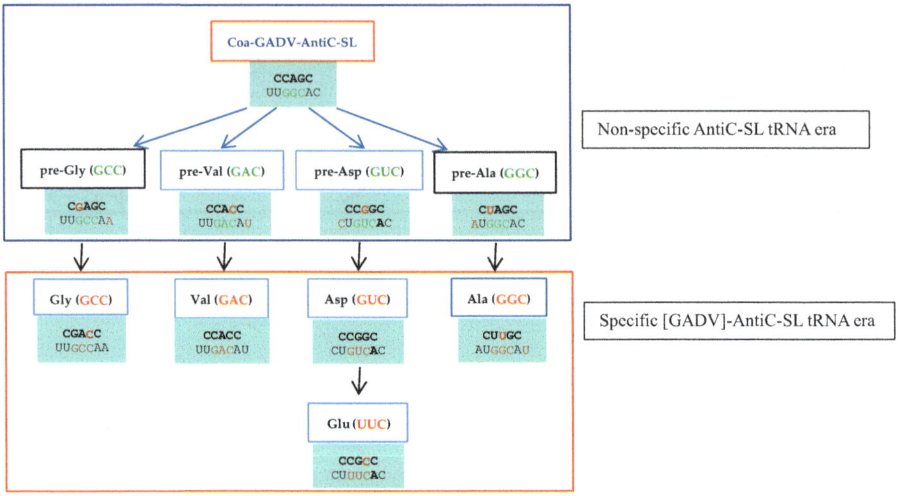

Fig. 6.8 The origin of ([GADV]+[E])-AntiC-SL tRNAs and evolutionary pathway from the first nonspecific to the five specific AntiC-SLs for translating GNC or GAR codon into five amino acids ([GADV]-amino acids and Glu [E]), respectively. It was found from analysis of 5′ stem sequences of AntiC-SL of *P. aeruginosa* tRNAs that the five specific AntiC-SLs (in red square) originated from one nonspecific AntiC-SL through four nonspedific AntiC-SLs (in blue square). Upper and lower base sequences indicate 5′ anticodon stem and anticodon loop sequences, respectively

between the AntiC-stem sequence and the pri-ARS recognizing the AntiC-stem sequence (Fig. 6.9).

Thus, the new origin and evolutionary pathway of the AntiC-SLs could be proposed, which are reasonable judging from several viewpoints as described below.

1. After the first nonspecific Coa-[GADV]-AntiC-SL was produced, four nonspecific precursor Gly (GCC)-, Ala (GGC)-, Asp (GTC)- and Val (GAC)-AntiC-SLs, were formed from the first nonspecific Coa-[GADV]-AntiC-SL through gene duplication of the tRNA gene. The four nonspecific AntiC-SLs further evolved into four specific Gly (GCC)-, Ala (GGC)-, Asp (GTC)- and Val (GAC)-AntiC-SLs, respectively (Figs. 6.8 and 6.9).
2. The first protein-synthesizing system could be formed with four descendant nonspecific [GADV]-AntiC-SLs before creation of the first gene (Fig. 6.9 (2)). In this way, formation of the four AntiC-SL tRNAs made it possible to form lateral dimer AntiC-SL tRNAs stimulating [GADV]-protein synthesis (Fig. 6.9 (2)).
3. More advanced protein-synthesizing system was further established using two pairs of AntiC-SLs bound vertically through the complementary anticodons (Fig. 6.9 (2)). At that time, formation of two pairs of amino acids, both of which are distinctively different from each other, as Gly (turn/coil forming amino acid)-Ala (α-helix forming amino acid) and Val (hydrophobic β-sheet forming amino acid)-Asp (hydrophilic turn/coil-forming amino acid), could make it possible to

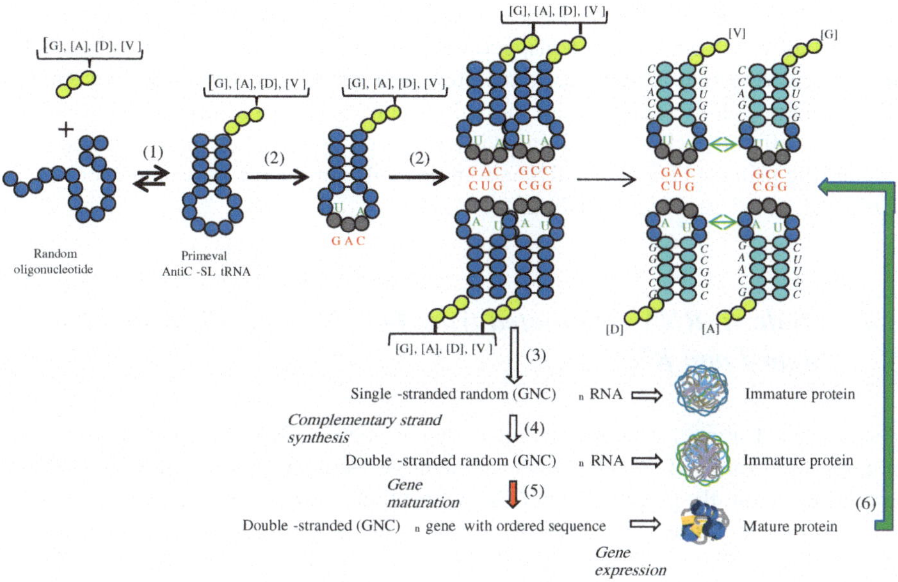

Fig. 6.9 Formation process of four specific AntiC-SL tRNAs for translating GNC codon. (1) The first primeval small nonspecific AntiC-SL, which can be randomly bound with one of [GADV]-amino acids at the 3′-terminal end, was formed through repeated random processes (see also Fig. 6.11). (2) Successively, four nonspecific AntiC-SLs carrying GNC anticodon were formed. The one AntiC-SL is shown in the figure. (3) The formation of four nonspecific AntiC-SLs triggered to creation of single-stranded (ss)-RNA with random $(GNC)_n$ sequence. (4) Furthermore, ds-$(GNC)_n$ RNA was created through complementary strand synthesis of the ss-RNA. Only immature water-soluble globular protein having random [GADV]-amino acid sequence could be produced just after creation of the ds-RNA because a random $(GNC)_n$ sequence was arranged on the both strands. (5) However, the ds-RNA could be matured to a gene encoding a mature protein like a precise polymer machine because of the double-stranded structure, with which base replacements can be memorized. (6) Appearance of mature [GADV]-ARS, which can discriminate one stem sequence from other stem sequences, made it possible to produce four AntiC-SLs with specific AntiC-stem sequence. These well indicate that AntiC-SL is the key for creation of the first $(GNC)_n$ gene and for synthesis of the first mature [GADV]-protein with ordered sequence

synthesize [GADV]-protein with a high catalytic activity, through incorporation of amino acids with strikingly different properties at a roughly equal probability.

4. Successively, the first ss-$(GNC)_n$ RNA was created by random joining of GNC anticodons, which were carried by the respective nonspecific [GADV]-AntiC-SLs (Fig. 6.9 (3)).
5. The creation of the first ss-$(GNC)_n$ RNA triggered formation of the first ds-$(GNC)_n$ RNA through complementary strand synthesis of the first ss-RNA (Fig. 6.9 (4)). The first ds-$(GNC)_n$ RNA gene encoding ARS, which can specifically recognize base sequences of AntiC-stem and anticodon in AntiC-loop, could be created for the first time through maturation of the first ds-$(GNC)_n$ RNA encoding nonspecific and immature ARS with a low catalytic activity (Fig. 6.9 (5)).

Thus, four specific [GADV]-AntiC-SL tRNAs could be formed through evolutionary pathway from the first nonspecific Coa-[GADV]-AntiC-SL tRNA (Fig. 6.8). It is noteworthy that anticodon, AntiC-stem and AntiC-loop of all the five AntiC-SLs contained in the network are mutually connected with a few members below one base difference (Fig. 6.8). It indicates that the vestiges showing formation process of the primeval [GADV]-tRNAs remains still now in the base sequences of 5′ AntiC-SLs of *P. aeruginosa* tRNAs.

3.4 Modern tRNA Originated from One Nonspecific Anticodon Stem-Loop RNA

As described above, it was concluded that modern tRNAs originated from one nonspecific AntiC-SL RNA. However, a large number of the AntiC-SL RNAs should be naturally prepared for [GADV]-protein synthesis. For example, a large number of Gly-AntiC-SL tRNA with the same length are required to produce [GADV]-protein. This is achieved only by expression of Gly-AntiC-SL tRNA gene. In addition, four kinds of AntiC-SL tRNAs with a same length must be also prepared for [GADV]-protein synthesis in order to polymerize four kinds of [GADV]-amino acids encoded by the GNC primeval genetic code.

Furthermore, amino group of one amino acid bound with one AntiC-SL tRNA must approach closely to carboxyl group of the other amino acid bound with the other AntiC-SL tRNA to make peptide bond between the two amino acid residues. For the purpose, two amino acids bound with two AntiC-SL tRNAs must be able to wobble not excess but sufficiently. The CCA terminal end makes it possible to give suitable mobility of amino acid bound with tRNA (Fig. 6.10). In the next section, it is discussed how the first AntiC-SL tRNA was created.

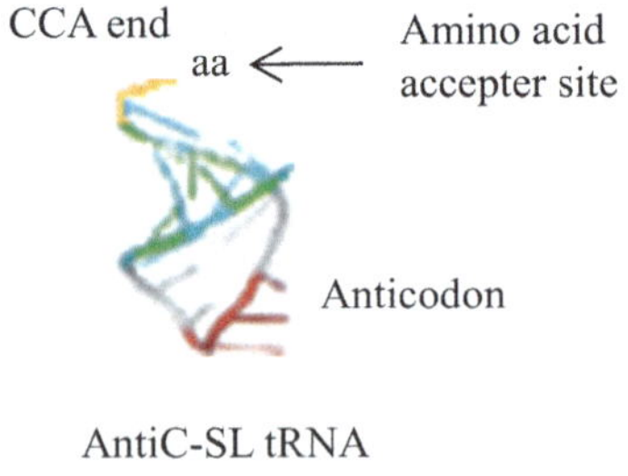

Fig. 6.10 The most primeval AntiC-SL structure assumed having CCA end, which can bind with an amino acid, and an anticodon in the anticodon loop. The whole structure is folded into a small and rigid hairpin-loop structure. The structure was drawn by combining two parts (CCA end and the AntiC-SL) of modern Yeast Phe-tRNA (PDBj, (1EHZ))

3.5 Properties of the First AntiC-SL tRNA

First, I present properties necessary to be the first primeval tRNA.

1. The first primeval tRNA must be created by random joining of monomeric units or nucleotides, only which could be carried out on the primitive Earth because of the absence of any tRNA and, therefore, the absence of any genetic function.
2. The first primeval single-stranded tRNA must be folded into a stable structure, which is resistant against hydrolysis, as easily understood the fact from that modern L-form tRNA is also stable, unlike mRNA (Fig. 6.10).
3. Even the primeval tRNA must have two indispensable abilities for tRNA. One is for reading a triplet codon sequence and the other is for carrying amino acid (Fig. 6.10).
4. A large number of the one kind of primeval tRNAs with the same length should be produced by using the first primeval tRNA gene, because it would be hard to create a plural number of other small and stable RNA structures with the same length through random processes.
5. The first primeval GNC genetic code is naturally realized with four kinds of primeval [GADV]-tRNAs carrying GNC anticodon. This can be confirmed from the results of analysis of 5′ stem sequences of *P. aeruginosa* tRNAs (Fig. 6.6), which is consistent with the GNC-SNS primitive genetic code hypothesis (Ikehara et al. 2002). In addition, the length of all the four kinds of primeval tRNAs must be the same, so that two amino acids bound with the two primeval tRNAs can stand close to form peptide bond between two amino acids easily.

These implies that gene for producing the first primeval tRNA should be formed when the first tRNA was created as shown in Fig. 6.11. Then, I confirmed whether or not the first primeval and other tRNAs produced according to the AntiC-SL hypothesis satisfy the above five properties. It can be supposed from the AntiC-SL hypothesis that the first primeval AntiC-SL tRNA could be formed as a small and stable stem-loop structure during repeated reaction cycles (2)~(4) starting from random joining of mononucleotides. Unstable RNA with a smaller number of base-pairs than others would be degraded at step (2). At step (3), complementary strand of undegraded RNA was synthesized to produce ds-RNA. The ds-RNA became the first tRNA gene. At step (4), single-stranded proto-primeval tRNA could be created by transcription of the ds-RNA (Fig. 6.11).

The AntiC-SL, which was finally obtained through the repeated reaction cycles (Fig. 6.11), should be naturally stable against hydrolysis. This is also supported by the fact that L-formed tRNA carrying AntiC-SL exposed to water is stable against hydrolysis by RNase (Fig. 6.1). However, many readers of this book may wonder whether such AntiC-SL tRNAs could be really produced by the repeated multi-step reaction cycles. I can answer about the concerns as follows. (1) According to the research carried out by Van der Gulik et al. (2009), it is known that [GADV]-amino acids are arranged at catalytic centers of enzymes, which catalyze RNA synthesis and degradation, at a high probability. (2) Furthermore, it is also known that GNC/

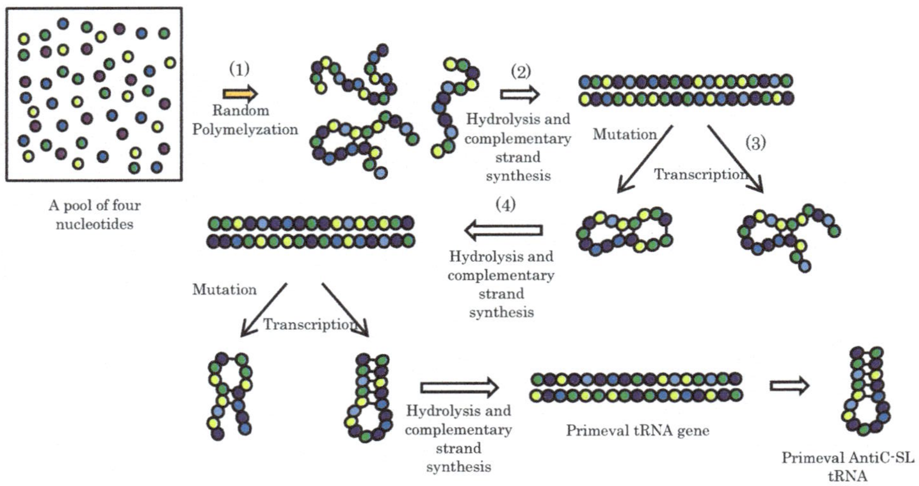

Fig. 6.11 A small and stable hairpin-loop RNA could be produced by (1) random joining of mononucleotides shown with four colored small circles, (2) degradation of unstable RNA with a smaller number of base pairs than others, followed by complementary strand synthesis, and (3) error-prone transcription by immature [GADV]-protein, which was produced by random joining of [GADV]-amino acids. After steps (2) and (3) were repeated, a sufficiently stable and the smallest hairpin-loop RNA could be finally obtained. The hairpin-loop RNA became the first AntiC-SL tRNA. As can be seen in the figure, the first primeval AntiC-SL tRNA gene was created when the first AntiC-SL tRNA was created in parallel

CNG base pairs are the most stable triplet bases among pairs formed between triplet bases (Taghavi et al. 2017). (3) Bases of anticodon loop of *P. aeruginosa* tRNAs translation GNC codons are not chemically modified unlike the cases of other *P. aeruginosa* tRNAs (tRNAdb (Universitat Leipzig)). These also imply that only AntiC-SL tRNAs carrying GNC anticodon can be folded into a stable hairpin-loop structure.

4 Evolutionary Process Deduced from Anticodon Stem-Loop Hypothesis

4.1 Evolutionary Pathway of tRNA Deduced from Mutual Relation Network of 5′ AntiC-Stem Sequences

Evolutionary pathway of tRNA can be interpreted from the whole network obtained with analysis of 5′ AntiC-stem sequences. The results are given in Fig. 6.12, indicating that the first primeval tRNA was an AntiC-SL translating G-start codon, actually one nonspecific AntiC-SL tRNA (Fig. 6.8), and that the genetic code or modern tRNAs originated from the primeval nonspecific AntiC-SL tRNAs for GNC codon encoding one of [GADV]-amino acids. Roughly speaking, the AntiC-SL tRNAs were evolved in order of the primitive tRNAs for translating G-start, C-start, A-start and U-start codons, because a hierarchy can be seen in the mutual relation map, suggesting that four AntiC-SLs for C-start codon (Leu (GAG1), His (GTG), Pro (GGG) and Pro (TGG)) were derived from previously existed four AntiC-SLs (Val (GAC), Asp (GTC), Ala (GGC) and Ala (TGC)), respectively (Figs. 6.6 and 6.12).

The reason why the respective AntiC-SL tRNA groups roughly form a cluster every N-start codon in the whole network would be because the genetic code evolved from G-start codons in order of to C-start codons, A-start codons and U-start codons. In other words, protein 0th-order structures are written on a single or four rows of the genetic code table, as [GADV]-amino acids encoded by GNC codon and SNS-encoding ten amino acids. S means G or C. Therefore, formation of new AntiC-SL tRNAs should generally proceed every row of the table in order to avoid duplicate usage of amino acids with similar physicochemical properties in every row or every N-start codon, so that water-soluble globular protein can be produced at a high probability under the newly evolved genetic code.

The evolutionary pathway is consistent with the direction assumed by GNC-SNS primitive genetic code hypothesis, suggesting that the genetic code originated from GNC primeval code, which evolved into the universal genetic code through SNS code (Ikehara et al. 2002). Therefore, it can be concluded that the mutual relation map given in Fig. 6.12 rightly represents the origin and evolutionary process of tRNAs, which originated from the primeval [GADV]-AntiC-SL tRNAs. Based on the results, I named the idea on the origin of tRNA described in this chapter as anticodon-stem loop (AntiC-SL) hypothesis (Ikehara 2019).

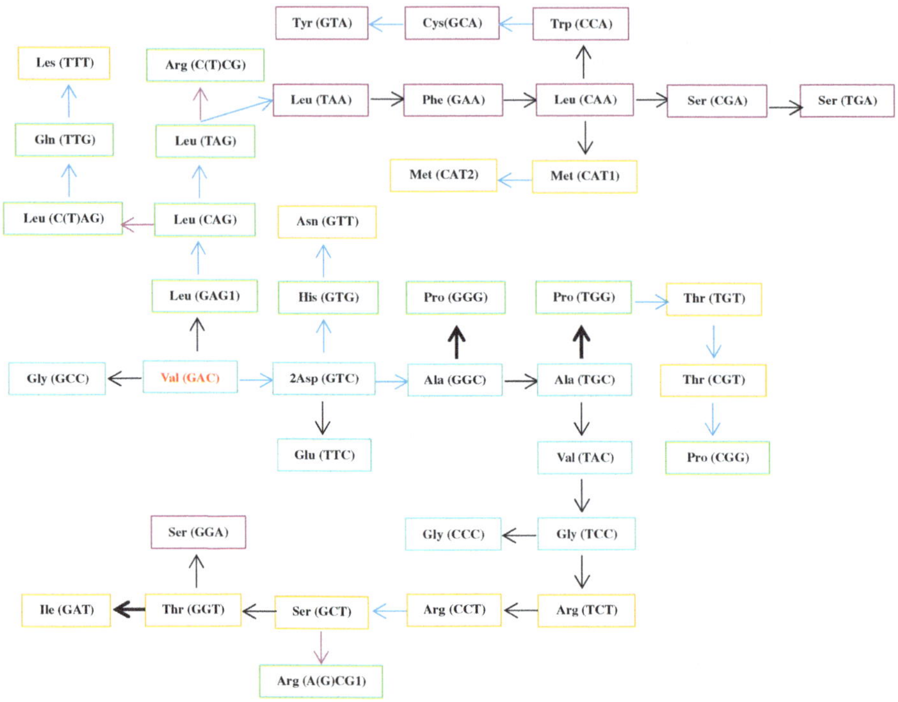

Fig. 6.12 A whole evolutionary relation network among forty-two, actually forty, because Leu (GAG2) and Arg (ACG2) do not have meaningful relation with any other *P. aeruginosa* tRNA 5′ AntiC-stem sequences and, therefore, cannot be integrated in the evolutionary relation map. The numeral 2 in the parenthesis of Leu (GAG2) and Arg (ACG2) means the second Leu (GAG) and Arg (ACG) tRNA. Amino acid name (anticodon) written in light blue, yellowish green, orange and pinkish purple boxes indicates 5′ AntiC-stem sequence of tRNA for G-, C-, A- and U-start codon, respectively. The map is drawn by connecting two AntiC-SL tRNAs, with single headed arrow, which indicates the evolutionary direction when it is assumed that AntiC-SL tRNAs originated from the hypothetical first ancestor Val-AntiC-SL tRNA written with red letters. Black bold, black thin and light blue arrows show strength of an evolutionary relation, which is connected with the number of conserved bases between the stem sequences of five, four and three out of the five bases, respectively. Pinkish arrow shows a special relation, such as Arg (ACG), Arg (CCG) and Leu (CAG), which is connected with other 5′ AntiC-stem sequence after base change from exceptional usage of A or C at the first base position of anticodon to more usual base of G or T in the parenthesis, respectively, in order to integrate them into the map

4.2 Evolutionary Process from the Primitive AntiC-SL tRNA to L-form tRNA

It is also important to confirm how the first AntiC-SL tRNA evolved to modern L-form tRNA, because the evolutionary process should further support the AntiC-SL hypothesis on the origin of tRNA, if there is no large irrational step in the evolutionary process.

It is well known that modern tRNA composed of around 75 nucleotides is folded into L-form. The CCA terminus carrying amino acid and anticodon in the anticodon loop are located at the two opposite ends of the L-form tRNA, as the former is at 3′ terminal end of accepter stem and the latter is at the other end of the L-form tRNA (Fig. 6.13b). Therefore, the distance from anticodon to amino acid bound at 3′-end of AntiC-SL tRNA is largely different from that of modern L-form tRNA (Fig. 6.13a and b). However, peptide bond must be formed between two amino acids bound with two tRNAs on the two sequential evolutionary steps from AntiC-SL to L-form tRNA, if the "anticodon-stem loop hypothesis" is correct. For example, the distance from anticodon to amino acid must be similar or in a permissible range to be able to make peptide bond between two amino acids bound with the AntiC-SL tRNA and one step evolved tRNA. Inversely stating this, it means that a strong evidence for the hypothesis should be obtained, if evolutionary process from the first AntiC-SL tRNA to L-form tRNA can be explained without any large contradiction at the respective evolutionary steps. Therefore, it is quite important to investigate the evolutionary process from the first AntiC-SL tRNA (Fig. 6.13a) to L-form tRNA (Fig. 6.13b).

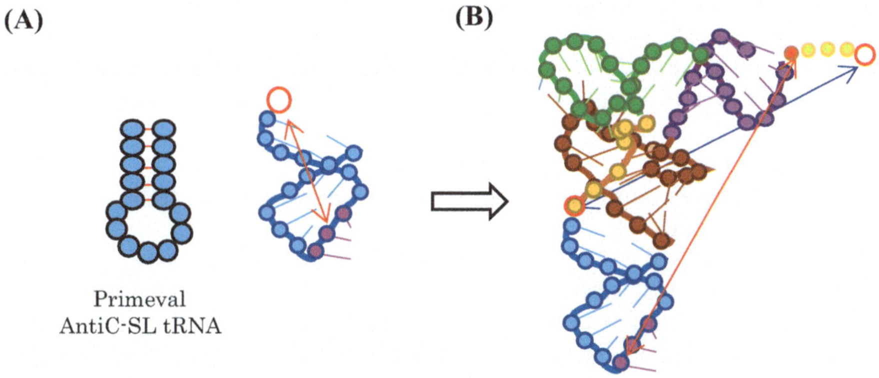

Fig. 6.13 (**a**) Structure of AntiC-SL deduced from the AntiC-SL hypothesis on the origin of tRNA. (**b**) Structure of L-form tRNA drawn as tracing yeast phenylalanyl-tRNA ((PDBj, (1EHZ)). As can be seen in the Figure, the distance from anticodon to 3′-end carrying amino acid is largely different between the two tRNAs, one is AntiC-SL (A) and L-form tRNA (B). Therefore, it is necessary to confirm whether or not there is no large discrepancy about the distance on the evolutionary process, further, whether or not genetic sequence could be translated by evolving tRNAs even at every evolutionary step from the first AntiC-SL tRNA (A) to modern L-form tRNA (B)

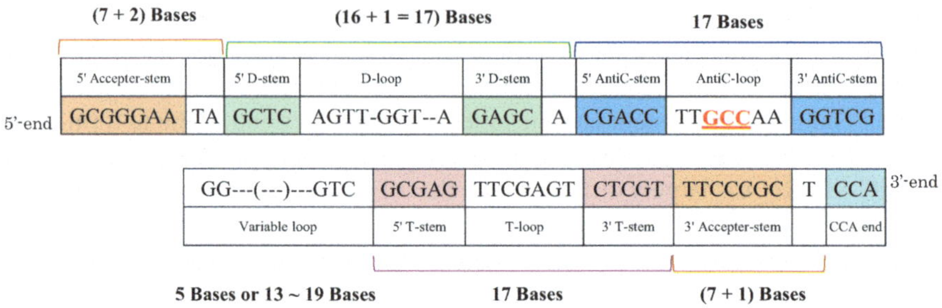

Fig. 6.14 Base sequence of *P. aeruginosa* Gly-tRNA (tRNAdb (Universitat Leipzig)). The sequences can be divided into four parts (17 bases long), except variable loop (5 bases long) and 3′ CCA end

4.2.1 L-form tRNA Was Created by Combination of Four AntiC-SL tRNAs

Modern tRNA is composed of three stem-loops, AntiC-SL, D-SL and T-SL, one accepter stem and one variable loop, except CCA end at the 3′ terminus (Fig. 6.14). The tRNA, for example in the case of *P. aeruginosa* Gly-tRNA, can be divided into the four parts, three stem loops and one accepter stem, all of which are composed of 17 nucleotides, except variable loop with 5 bases long (Fig. 6.14). This suggests that two stem-loops of D-SL and T-SL and one accepter stem was derived from the most primitive, one AntiC-SL tRNA. Therefore, it is assumed that the modern L-form tRNA was formed by adding one AntiC-SL to the AntiC-SL tRNA or an intermediate tRNA, which was formed on the way of evolution from the first AntiC-SL to L-form tRNA. On the contrary, it is assumed that variable loop was added through a quite different pathway from the formation process of D-SL, T-SL or accepter stem.

4.2.2 Deduced Evolutionary Steps from the First AntiC-SL tRNA to L-form tRNA

Here, it is assumed as a premise that three parts were added to the AntiC-SL or intermediates previously formed, to create the L-form tRNA at minimum steps. Therefore, the formation process of the L-form tRNA was explored, assuming that a stem-loop structure, which was once added to an intermediate, did not move to another place and that the three parts, which were originally AntiC-SL, were piled up to the first AntiC-SL or intermediates previously formed one by one. Under the conditions, it was confirmed whether or not a large contradiction was not caused during integrating the three parts into the first AntiC-SL.

Of course, there are four possibilities considering that one of the four parts including variable loop were added to the first AntiC-SL tRNA at the first step. Among the four possibilities, I selected the evolutionary process in which the four parts were added to the most primitive AntiC-SL tRNA in order of accepter stem, D-SL, variable loop and T-SL, because both AntiC-SL and accepter stem should be stable against degradation by RNase, judging from the fact that the two structures of modern L-form tRNA are exposed to water (Fig. 6.1) and also because any large discrepancy did not occur, as explained below. An evolutionary process from the most primitive AntiC-SL tRNA to L-form tRNA could be reasonably explained as a result.

(1) The first step: Addition of one AntiC-SL to the other AntiC-SL to form accepter stem

The stem end of one AntiC-SL was inserted into the loop of the other AntiC-SL to form (AntiC-SL+ accepter stem) tRNA (Fig. 6.15a). The reasons are because both CCA-end (amino acid accepter site) and discrimination sequence of the most primitive AntiC-SL tRNA could be moved to the accepter stem (so called "the second code") of the new tRNA without interruption.

However, it seems that the two distances between anticodon and amino acid accepter site of AntiC-SL tRNA and between those of (AntiC-SL+ accepter stem) tRNA are apparently inconsistent (Fig. 6.15 (A)). However, it could be confirmed that there is not any large discrepancy between the two distances from anticodon to amino acid accepter site because of unique ds-RNA structure (Fig. 6.15b).

Assume here that the accepter stem were formed independently of the primitive AntiC-SL by combining two oligonucleotide composed of 7 nucleotides. However, it would be difficult to form two oligonucleotides having a complementary sequence each other, because of the small appearance probability ($4^7 = {\sim}10^{-4}$). This also supports the idea that one AntiC-SL was added to the other AntiC-SL to form (AntiC-SL +accepter stem) at the first step.

(2) The second step: Insertion of another AntiC-SL between AntiC-SL and accepter stem to form D-SL

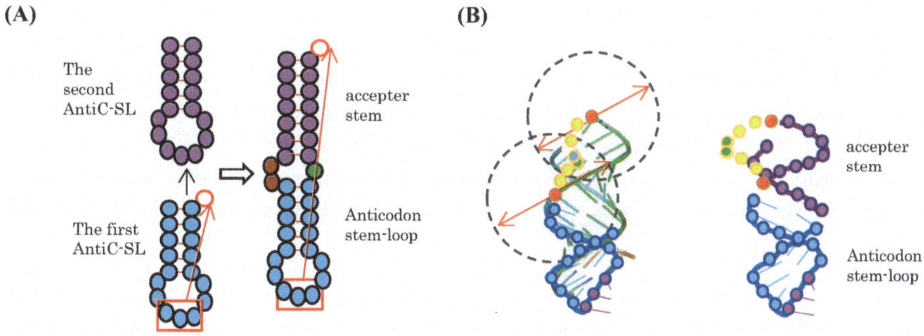

Fig. 6.15 The first evolutionary step from the first AntiC-SL tRNA to (AntiC-SL + accepter stem) tRNA. (**a**) It seems that distance (red arrow) between anticodon (red rectangle) and the amino acid accepter site (red circle) changes largely by addition of the second AntiC-SL onto the first AntiC-SL. (**b**) However, it could be confirmed that the change of the site (small green circles), where amino acid is bound, is within an allowable range, which is shown with double-headed arrows and broken circles, because of the characteristics of ds-RNA and the flexible single-stranded CCA end (three yellow circles)

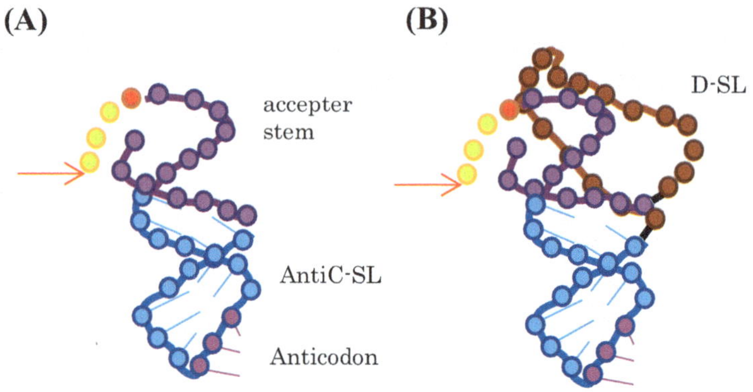

Fig. 6.16 The evolutionary step from the (AntiC-SL + accepter stem) tRNA to the (AntiC-SL + accepter stem + D-SL) tRNA. (**a**) The location of CCA-end of (AntiC-SL + accepter stem) tRNA is shown by a short red arrow. (**b**) The location of CCA-end of (AntiC-SL + accepter stem + D-SL) tRNA is also shown by a short red arrow. It can be seen that the location of CCA-end is almost unchanged by insertion of D-SL between AntiC-SL and accepter stem of (AntiC-SL + accepter stem) tRNA. Short black lines indicate bonds connecting between two stem-loop structures

It is considered that accepter stem was stabilized by adding the third AntiC-SL to the (AntiC-SL + accepter stem). The location of amino acid bound at the CCA-end was almost unchanged before and after the addition of D-SL (Fig. 6.16a and b). Here, D-SL of *P. aeruginosa* Phe-tRNA was used instead of the AntiC-SL, because it is supposed that amino acid sequence and structure of the original AntiC-SL largely changed to optimal sequence and structure of the modern D-SL during the evolution to L-form tRNA.

(3) The third and the fourth steps: Addition of variable loop to the (AntiC-SL + accepter stem + D-SL) tRNA followed by integration of T-SL. A few nucleotides of variable loop were added to the (AntiC-SL + accepter stem + D-SL) tRNA one by one as adjusting the amino acid binding site at CCA-end (Fig. 6.17a and b). T-SL was further inserted between 3′ end of growing variable loop and 5′ end of one strand of accepter stem (Fig. 6.17b). Thereafter, several nucleotides of variable loop were further added between 5′ end of T-SL and 3′end of growing variable loop one by one (Fig. 6.17c). Therefore, it is assumed that variable loop played a role in adjusting two locations between anticodon and 3′-CCA-5′ end of evolving tRNA.

Thus, it was found that evolutionary process from the first AntiC-SL to L-form tRNA can be explained without any large inconsistency during the evolution. This also supports the idea assuming that the modern L-form tRNA originated from the one primitive AntiC-SL tRNA.

Strengths of AntiC-SL tRNA Hypothesis It could be concluded according to the necessary conditions for the origin of something that AntiC-SL tRNA hypothesis is valid as the origin of tRNA, because both formation process of the most primitive

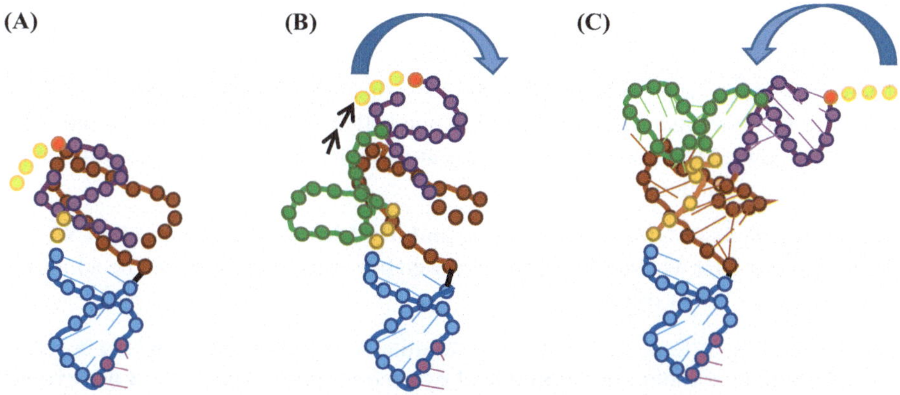

Fig. 6.17 The evolutionary step from the (AntiC-SL + accepter stem + D-SL) tRNA to the L-form tRNA. (**a**) Two nucleotides (small light brawn circles) of variable loop were added into (AntiC-SL + accepter stem + D-SL) tRNA. (**b**) T-SL was inserted at between 3′ end of growing variable loop composed of three nucleotides and 5′ end of one strand of accepter stem of (AntiC-SL + accepter stem + D-SL) tRNA. Short black arrows indicate rotation of (T-stem loop + accepter stem) accompanied by some nucleotide additions of variable loop. (**c**) The final structure of L-form tRNA after addition of all nucleotides of variable loop. Note that location of amino acid bound with 3′ CCA end could be adjusted by swinging of single-stranded CCA (shown by bold curved arrows) every one nucleotide addition of variable loop

AntiC-SL tRNA through random process and evolutionary process from the AntiC-SL tRNA to modern L-form tRNA can be explained without any large contradiction.

Weaknesses of AntiC-SL tRNA Hypothesis As described in the above strengths of AntiC-SL tRNA hypothesis, I have considered that the most primitive tRNA was AntiC-SL tRNA as supposed by the hypothesis. However, simultaneously, I have considered that there is a great weakness in the hypothesis that several matters assumed in the hypothesis have not been confirmed by experiments as described below.

1. Could AntiC-SL RNA be really produced through the synthesis-degradation cycles of oligonucleotides with immature [GADV]-proteins?
2. Could modern L-form tRNA be really formed by addition of four AntiC-SLs as expected by the AntiC-SL tRNA hypothesis? and so on.

5 Discussion

There are at least two conditions for admitting that something assumed is the most primitive form of modern tRNA. The first one is that the most primitive something must be produced through random process and the second one is that the something could evolve to modern L-form tRNA.

Validity of Ideas of Other Researchers and My Idea on the Origin of tRNA
Then, first consider whether the ideas proposed by other researchers can be admitted as the origin of tRNA or not.

(1) It would be difficult to consider the minihelix as the origin of tRNA, because the minihelix could not be produced by random process and could not evolve into modern L-form tRNA (Fig. 6.18a). In addition, anticodon should appear for the first time, when RNA composed of D-stem loop and AntiC-SL was bound with the minihelix composed of T-stem loop and accepter stem, according to the minihelix hypothesis. Further, any genetic information should not be read by the minihelix tRNA and the translation system could not be transferred from a system without using anticodon to another system with anticodon, when the RNA composed of D-stem loop and AntiC-SL was added to the minihelix. However, it would be also quite difficult to rationally explain the transition process.
(2) Modern tRNA would not originated from the double-hairpin RNA, because it would be also difficult to produce the two double-hairpins through random process (Fig. 6.18b). The reasons are as follows.

1. It is assumed by the double-hairpin hypothesis that two halves of modern tRNA were first formed. However, it would be difficult to produce the two halves through random process, because the appearance probability should be too small as $1/4^{30/2} = \sim 1/10^9$ to produce the two halves through random process at one stroke.
2. In addition, it would be probably impossible to convert from a translation system using double-hairpin tRNA to modern translation system carried out with L-form tRNA, because it is necessary to convert the double-hairpin RNA to L-form tRNA at one stroke.

Next, consider whether or not the AntiC-SL hypothesis, which I have proposed, is valid for the origin of tRNA.

For the purpose, I first enumerate the conditions, which are necessary to confirm that the AntiC-SL is suitable to the origin of tRNA.

1. The first primitive tRNA must be produced through random process.
   It is supposed that the first AntiC-SL tRNA was created through repeated random synthesis and degradation of RNA, as shown in Fig. 6.11.
2. The most primitive tRNA must have features similar to the modern one.
   The first AntiC-SL tRNA was made of single-stranded RNA, which was folded into a small and stable hairpin-loop RNA and therefore the first AntiC-SL tRNA had basic features similar to modern tRNA from the beginning. (Fig. 6.10).

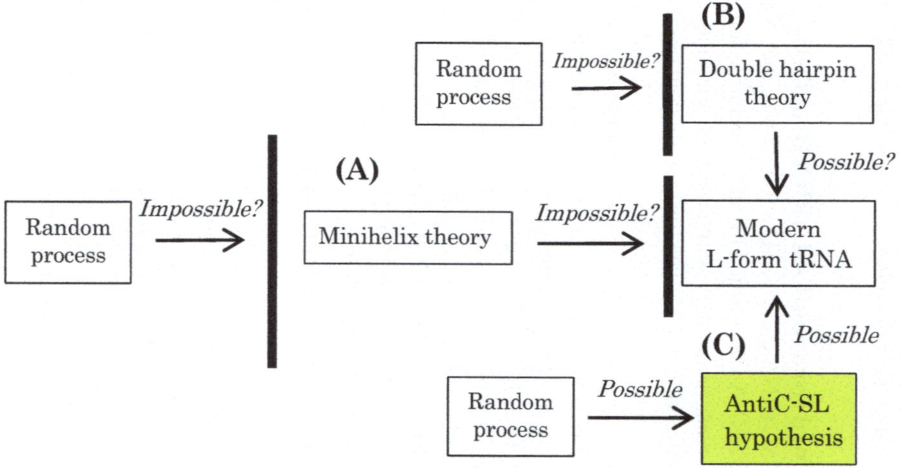

Fig. 6.18 (**a**) and (**b**) Sizes of primitive tRNAs assumed by minihelix theory (T-stem loop + accepter stem) and double hairpin theory (two halves of a whole tRNA) are too large to form through the respective random processes. In addition, it would be probably impossible to form modern L-form tRNA from the minihelix. (**c**) On the contrary, the AntiC-SL, which is proposed as the most primitive tRNA in the AntiC-SL tRNA hypothesis, could be formed through random process and also could evolve to modern L-form tRNA as explained in the text

3. The most primitive tRNA assumed must have at least both anticodon and accepter site of amino acid.

   The first AntiC-SL tRNA have both anticodon and CCA end accepting amino acid (Fig. 6.10).

4. The most primitive tRNA assumed must be able to evolve to the modern one.

The first AntiC-SL tRNA could evolve to the modern L-form tRNA as explained in Sect. 4.2.

As described above, the AntiC-SL tRNA satisfies all the four conditions (see also Fig. 6.18c). Therefore, it could be concluded that the AntiC-SL tRNA was the first tRNA.

The Essence of tRNA Viewed from AntiC-SL tRNA Hypothesis

Why could the first AntiC-SL tRNA carry GNC anticodon in the loop? The reason is not because the AntiC-SL tRNA was formed to carry the anticodon in the loop but because GNC anticodon was fortunately contained in the anticodon loop of AntiC-SL tRNA. That is, anticodon, GNC, was fortunately contained in a small and stable hairpin-loop RNA, which was accidentally generated and selected during repeated RNA synthesis and its degradation cycles. In addition, the generation of the AntiC-SL tRNA could also fortunately lead to formation of the first $(GNC)_n$ gene.

Similarly, the reason, why the first genetic code was GNC triplet code, is not because the GNC code was used to be able to encode 20 amino acids in future. Not the case, the reason, why 20 amino acids could be fortunately encoded by the universal genetic code, would be simply because the triplet GNC code, which was included in the loop of the first AntiC-SL tRNAs, was accidentally used as the first genetic code, which can develop to the universal genetic code encoding 20 amino acids, as if it were prepared from the beginning.

Why Are the Four Nucleobases (A, U, G, C) Used as Genetic Materials?

One more important question is why the four nucleobases (A, U, G, C) or four nucleotides are used as components of genetic information carriers. Although I do not know the exact reason at the present time point, I can state at least, the reason would be because the four AntiC-SL RNAs, with which two sets of triplet GNC/CNG base pairs can be formed, could be fortunately produced in the absence of gene. Consequently, life could emerge on this planet owing to the use of the AntiC-SL tRNAs, which were composed of four nucleotides. The use of the AntiC-SL tRNAs would lead to the use of RNA/DNA as genetic information carriers.

6 Conclusion

Transfer RNA or tRNA including the most primitive AntiC-SL tRNA can be regarded as a mediator of genetic information (codon) to amino acid sequence of protein (amino acid). Therefore, the essence of the tRNA is described as follows.

1. tRNA must have anticodon reading triplet codon for expression of genetic information.
2. In addition, tRNA must carry amino acid binding site for protein synthesis.

Therefore, both modern L-form tRNA, which should originate from one small and stable AntiC-SL tRNA, and the first AntiC-SL tRNA have and had both an anticodon and CCA 3′-end as a binding site of amino acid. Consequently, the most primitive triplet GNC codon carried by the AntiC-SL tRNA determined a frame of the modern universal genetic code and also the AntiC-SL tRNA played a key role in creation of genetic sequence as acquiring support from protein 0th-order structure or [GADV]-amino acids.

Therefore, it can be stated that tRNA but not mRNA is unquestionably one of the important members in the core life system together with gene and protein (Chap. 2: Fig. 2.1).

References

Di Giulio M (2006) The non-monophyletic origin of the tRNA molecule and the origin of genes only after the evolutionary stage of the last universal common ancestor (LUCA). J Theor Biol 240:343–352

Di Giulio M (2008) The split genes of Nanoarchaeum equitans are an ancestral character. Gene 421:20–26

Fujishima K, Kanai A (2014) tRNA gene diversity in the three domains of life. Front Genet 5:142

Ikehara K (2019) The origin of tRNA deduced from *Pseudomonas aeruginosa 5′* anticodon-stem sequence: Anticodon stemloop hypothesis. Orig Life Evol Biosph 49:61–75

Ikehara K, Omori Y, Arai R et al (2002) A novel theory on the origin of the genetic code: a GNC-SNS hypothesis. J Mol Evol 54:530–538

Kim SH, Quigley G, Suddath FL et al (1972) The three-dimensional structure of yeast phenylalanine transfer RNA: shape of the molecule at 5.5-A resolution. Proc Natl Acad Sci U S A 69:3746–3750

Rodin S, Rodin A, Ohno S (1996) The presence of codon-anticodon pairs in the acceptor stem of tRNAs. Proc Natl Acad Sci U S A 93:4537–4542

Root-Bernstein R, Kim Y, Sanjay A et al (2016) tRNA evolution from the proto-tRNA minihelix world. Transcription 7:153–163

Schimmel P, de Pouplana LR (1995) Transfer RNA: from minihelix to genetic code. Cell 81:983–986

Schimmel P, Giegé R, Moras D (1993) An operational RNA code for amino acids and possible relationship to genetic code. Proc Natl Acad Sci U S A 90:8763–8768

Sugahara J, Fujishima K, Morita K et al (2009) Disrupted tRNA gene diversity and possible evolutionary scenarios. J Mol Evol 69:497–504

Taghavi A, van der Schoot P, Berryman JT (2017) DNA partitions into triplets under tension in the presence of organic cations, with sequence evolutionary age predicting the stability of the triplet phase. Q Rev Biophys:e15. https://doi.org/10.1017/S0033583517000130

Tamura K (2015) Origins and early evolution of the tRNA molecule. Life (Basel) 5:1687–1699

Tamura K, Schimmel PR (2006) Chiral-selective aminoacylation of an RNA minihelix: mechanistic features and chiral suppression. Proc Natl Acad Sci U S A 103:13750–13752

Tanaka T, Kikuchi Y (2001) Origin of the cloverleaf shape of transfer RNA-the double-hairpin model: implication for the role of tRNA intron and the long extra loop. Viva Orig 29:134–142

Van der Gulik P, Massar S, Gilis D et al (2009) The first peptides: The evolutionary transition between prebiotic amino acids and early proteins. J Theor Biol 261:531–539

Chapter 7
The Origin of the Genetic Code

Abstract Genetic code mediating between a codon sequence of a gene and an amino acid sequence of a protein is one of the core members of modern genetic system together with gene and protein. Therefore, many researchers have tried to elucidate the origin of the genetic code. The origin of the genetic code must be considered as dividing two sides. That is, the first one is from what kind of code the genetic code originated and the second one is how correspondences between a codon and an amino acid were determined. About the first problem, several hypotheses, as RNY hypothesis, GC code hypothesis, GCU hypothesis and so on, have been proposed by other researchers thus far. However, the three hypotheses have been presented mainly based on the appearance frequency of codons in modern genes. Another hypothesis, the four column theory, has been presented based on the amounts of amino acids produced with prebiotic means. Therefore, all the four hypotheses are not based on protein structure formation, which should naturally relate to the origin of the genetic code. Therefore, the hypotheses proposed by other researchers cannot explain rationally the origin and evolution of the genetic code. On the other hand, GNC-SNS primitive genetic code hypothesis, which was proposed chiefly based on the protein structure formation, can well explain the origin and evolution of the genetic code. About the second one for explaining the establishment process of the genetic code or the correspondences between a codon and an amino acid, two theories have been mainly proposed thus far. One is the stereochemical theory and the other is the frozen-accident theory. However, there are critical defects in both theories, with which the whole correspondences cannot be rationally explained. On the other hand, I recently proposed the anticodon stem-loop (AntiC-SL) tRNA hypothesis, assuming that modern tRNAs originated from one nonspecific AntiC-SL tRNA through four [GADV]-AntiC-SL tRNAs. When I considered the process how the correspondences between four GNC codons and four [GADV]-amino acids were established, a new idea suddenly came to my mind. The idea is "GNC code frozen-accident theory", considering that the correspondence relations between GNC codons and [GADV]-amino acids were determined accidentally and were frozen. Thereafter, new amino acids were captured into the first GNC genetic code one by one, presumably in order of Glu, Leu, Pro, His and so on, to complement insufficient abilities of the proteins composed of amino acids encoded by the previous genetic code. The origin and evolutionary process of the genetic code deduced from the GNC code frozen-accident theory are consistent

K. Ikehara, *Towards Revealing the Origin of Life*,
https://doi.org/10.1007/978-3-030-71087-3_7

with GNC-SNS primitive genetic code hypothesis, with both the coevolution theory and the adaptive theory on the origin and evolution of the genetic code, and even with the GADV hypothesis on the origin of life. The "GNC code frozen-accident theory" has been presented in this Chapter. The origin and the whole evolutionary process of the genetic code could be reasonably explained with the GNC code frozen-accident theory and GNC-SNS primitive genetic code hypothesis.

Keywords Origin of genetic code · RNY hypothesis · GC hypothesis · Four column theory · GCU code hypothesis · Stereochemical theory · Frozen accident theory · GNC-SNS primitive genetic code hypothesis · GNC code · SNS code · GNC code frozen-accident theory · Coevolution theory · Adaptive theory

Marshall Nirenberg, who deciphered the genetic code, gave a lecture in Concluding Remarks of his Nobel Lecture, December 12, 1968 as follows.

> *The genetic code is now essentially deciphered. I have been fortunate in having the collaboration of many enthusiastic associates during the course of our studies. To do justice to the years of effort and the important contributions made by associates and numerous colleagues throughout the world is virtually impossible in the available time. One has only to refer to the comprehensive reviews in the Cold Spring Harbor Symposium on Quantitative Biology of 1963 and 1966 to view the breadth of the field and the extent of information now available.*

The universal or standard genetic code, which was deciphered by Nirenberg et al. is amazingly constructed. In this Chapter, it is answered to the questions; from what code the modern genetic code originated, how the correspondences between codon and amino acid were established, and how the most primitive genetic code evolved to the modern genetic code (Figs. 7.1 and 7.2).

1 Introduction: Towards Solving the Riddle of the Origin of the Genetic Code

There are two sides for exploring the origin of the genetic code. The first one is to understand from what primitive genetic code the modern one originated and how the primitive genetic code has evolved to the modern genetic code. The second one is to clarify how the correspondences between a codon and an amino acid were determined. On the other hand, the genetic code is actually realized by tRNA, which determines the corresponding relation between a codon and an amino acid. Therefore, the genetic code is a table summarizing the correspondences, which were determined by tRNAs. Based on the origin of tRNA, it is discussed whether hypotheses on the origin of the genetic code, which have been proposed by other researchers and I have considered, are valid or not. For achieving the purposes, it is indispensable to well understand a final goal toward which the most primitive genetic code evolved.

Fig. 7.1 Universal (standard) genetic code table. From deciphering the genetic code, it was found that the genetic code has been beautifully constructed, for example, as arranging hydrophobic amino acids at the most left column. Instead, the reasons, why the genetic code is formed so beautifully and how was the first genetic code created through random process on the primitive Earth, remain unexplained. The answers for solving the riddle of the origin of genetic code are given in this Chapter

	U	C	A	G	
U	Phe	Ser	Tyr	Cys	U
	Phe	Ser	Tyr	Cys	C
	Leu	Ser	Term	Term	A
	Leu	Ser	Term	Trp	G
C	Leu	Pro	His	Arg	U
	Leu	Pro	His	Arg	C
	Leu	Pro	Gln	Arg	A
	Leu	Pro	Gln	Arg	G
A	Ile	Thr	Asn	Ser	U
	Ile	Thr	Asn	Ser	C
	Ile	Thr	Lys	Arg	A
	Met	Thr	Lys	Arg	G
G	Val	Ala	Asp	Gly	U
	Val	Ala	Asp	Gly	C
	Val	Ala	Glu	Gly	A
	Val	Ala	Glu	Gly	G

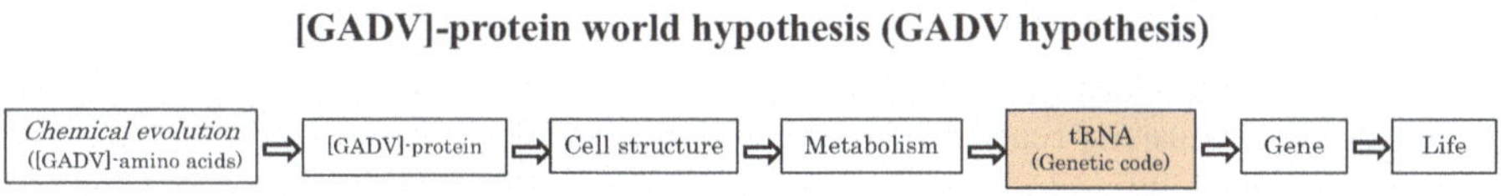

Fig. 7.2 In this Chapter, the origin of genetic code is discussed (red box). Establishment of the most primitive genetic code was triggered by formation of the most primitive anticodon stem-loop (AntiC-SL) tRNA (red box)

1.1 *Necessary Matters for Exploration of the Origin of the Genetic Code*

First, the matters, which should be kept in mind to explore the origin of the genetic code, are enumerated as follows.

1. It is necessary to well understand what features modern genetic code has, in order to make clear the origin of the genetic code, a member composing the fundamental life system.

2. Evolutionary process from the primitive genetic code, which should be created by random processes on the primitive Earth, to modern genetic code, must be explained reasonably.
3. If it was assumed that the contemporary genetic code originated from a primitive one with features different from the contemporary genetic code, as a matter of course, first, conversion process from the primitive genetic code to the most primitive one with similar features to the modern one must be reasonably explained, either directly or indirectly.

Furthermore, the following matters also must be considered and well understood.

1. As tRNA bridging over a codon and an amino acid has been formed to synthesize a protein, the genetic code is a mere table representing the correspondences between a codon and an amino acids, which have been realized by tRNA. Therefore, the genetic code never be created before formation of tRNA or independently of tRNA.
2. Every genetic code, which appeared during the evolutionary process from the first genetic code, should give a guarantee of production of a water-soluble globular protein by random joining of amino acids, which are written in the genetic code or in a part of the genetic code. This indicates that tRNA was first created to efficiently produce a protein and the genetic code was established as the results of the origin and evolutionary process of tRNA.
3. Therefore, it is important to understand that the origin and evolutionary process of the genetic code must be considered in relation to the protein structure formation in addition to the origin and evolution of tRNA.

Then, general features of the genetic code are enumerated as dividing two parts, general structural features and functional features as follows.

1.2 General Structural Features of the Genetic Code

1. The correspondences between a codon and an amino acid in the modern genetic system are indirectly realized through amino acid addition to 3′ end of tRNA and recognition of anticodon of tRNA by aminoacyl tRNA synthetase (ARS) recognizing both an amino acid and an anticodon (Chap. 3: Fig. 3.1).
2. Amino acids are excellently arranged in the genetic code table, as amino acids with similar physical and chemical properties are generally positioned in the same column (Fig. 7.1).
3. Amino acids with different physical and chemical properties are generally arranged in the same row.
4. The genetic code is degenerated as the same amino acid occupies a two or four codon box.

1.3 General Functional Features of the Genetic Code

1. The genetic code indicates the correspondence relation between a codon and an amino acid, which is necessary to translate genetic information into amino acid sequence.
2. The genetic code is constructed to produce a protein efficiently under a random process or even by random joining of amino acids in a protein 0th-order structure. Therefore, protein 0th-order structures are written in the genetic code table, so that protein can be produced effectively even under the random process.

1.4 There Are Two Sides for Exploring the Origin of the Genetic Code

1. From what code did the genetic code originate and how the first genetic code evolved to the modern universal genetic code?
2. How were the correspondences between a codon and an amino acid established?

Therefore, I discuss as dividing research purposes of the origin of the genetic code into the two in this Chapter.

2 The Ideas on the Origin of the Genetic Code Advocated by Other Researchers

2A. ***The Origin of the Genetic Code-1:*** **From What Code the Genetic Code Originated?**
As well known, the genetic code mediates a codon sequence of a gene with an amino acid sequence of a protein. Therefore, it is critically important to understand what kinds of amino acids were used in the first genetic code, because primeval water-soluble globular proteins should be formed by random joining of the amino acids encoded by the first genetic code before creation of the first gene. Several hypotheses on the origin of the genetic code-1 have been proposed by other researchers as follows.

2.1 RNY Hypothesis

The RNY hypothesis was proposed based on the facts that average base compositions at three codon positions of genes used in extant organisms are similar to RNY. R and Y mean purine bases, A and/or G, and pyrimidine bases, U (T) and/or

C. The hypothesis was described in the text book "Molecular biology of the gene, 4th-ed." of Watson, et al. (1987).

Strengths of RNY Hypothesis As well known, the genetic information for protein synthesis is written as codon sequence on DNA or RNA strand. Some hypotheses on the origin of the genetic code have been frequently proposed founded on the base compositions at the three codon positions. The RNY hypothesis would be a natural idea, which has been presented from the viewpoint or based on the usage frequency of codon.

Weaknesses of RNY Hypothesis However, the reason, why the base compositions at the three codon positions are roughly RNY, has not been explained in detail. On the contrary, the reason can be simply interpreted according to the GNC-SNS primitive genetic code hypothesis, which I have proposed, suggesting that the genetic code originated from GNC code. Therefore, the reason, why the frequency of usage of RNY is relatively high even in modern genes, can be reasonably explained as below. The properties, that G (one of purine bases: R) and C (one of pyrimidine bases: Y) were simply used at the first and the third codon positions of the most primitive genes, respectively, descended to many genes, which were used under a subsequent genetic code, SNS, and even under the universal genetic code.

2.2 GC Hypothesis

Hartman has proposed the GC code hypothesis, assuming that the genetic code originated from GC code, which is composed of 16 codons (GGN, GCN, CCN and CGN) encoding four amino acids, Gly, Ala, Pro and Diapr (L-2,3-diaminopropionic acid) instead of Arg (Table 7.1 (A)) (Hartman and Smith 2019).

Strengths of GC Hypothesis The GC code has been proposed based on the fact that many proteins use four amino acids (Gly, Ala, Pro and Arg (Diapr)) encoded by the GC code at a relatively high frequency. It is also a natural idea that Diapr is integrated into the four amino acids instead of Arg, because it is difficult to synthesize Arg with prebiotic means.

Weaknesses of GC Hypothesis However, there are some weaknesses in the hypothesis, as described below.

(1) It would be difficult to create the sixteen codons in the GC code encoding four amino acids (GGN for Gly, GCN for Ala, CCN for Pro and CGN for Diapr) at one stroke by random processes on the primitive Earth (Table 7.1 (A)).
(2) Non-natural Diapr is used in the GC code instead of Arg, which is difficult to produce with prebiotic means. Certainly, Diapr with only three carbon atoms should be easily synthesized and accumulated at a large amount on the primitive Earth. However, Diapr with a small and positively charged side chain

Table 7.1 **(A)** GC code hypothesis on the origin of the genetic code. Four amino acids, Gly, Ala, Diapr (L-2,3-diaminopropionic acid) and Pro, are encoded by sixteen SSN codons. Only one α-helix forming amino acid, Ala, and one hydrophilic amino acid, Diapr, are used and instead both β-sheet forming amino acid and hydrophobic amino acid are not contained in the GC-code. **(B)** In the GNC code, four [GADV]-amino acids, Gly, Ala, Asp and Val, are encoded by four GNC codons. Three secondary structure forming amino acids and both hydrophilic and hydrophobic amino acids are included in the four amino acids encoded by the GNC code. Hydrophilic and hydrophobic amino acids are written in blue and yellowish brown boxes, respectively. Symbols, + and -, indicate a high and a low hydropathy, respectively

(A) GC code

GGN	GCN	CGN	CCN
Gly	Ala	Diapr	Pro
Turn/coil	β-Helix	Turn/coil	Turn/coil
+/-	++/-	Hydrophilic	+/- -

(B) GNC code

GGC	GCC	GAC	GUC
Gly	Ala	Asp	Val
Turn/coil	α-Helix	Turn/coil	β-Sheet
+/-	++/-	Hydrophilic	Hydrophobic

($-CH_2--NH_2$) should be a highly turn/coil forming amino acid. Therefore, Diapr should not be used in the GC code in order to avoid overlapping use with Gly and Pro, both of which are turn/coil forming amino acids, although Diapr, of course, might be accidentally incorporated at a small probability into primitive [GADV]-proteins during random joining of [GADV]-amino acids before establishment of the first GNC genetic code (Table 7.1 (B)) (Ikehara et al. 2002).

(3) Glu, which is required to synthesize Pro as a precursor molecule in the modern metabolism, is not contained in the GC code. This is apparently incompatible with the co-evolution theory on the origin of the genetic code, suggesting that a precursor amino acid should be encoded in a genetic code before a product amino acid is incorporated into the genetic code (Wong 1975; Wong et al. 2016; Di Giulio 2008).

(4) Both an amino acid with a high β-sheet formability and a hydrophobic amino acid, which are generally indispensable to form a water-soluble globular protein, are not contained in the GC code.

(5) Physicochemical properties, for example, length and bulkyness of side chain, of Diapr are largely different from those of Arg. Therefore, it is supposed that deleterious multiple mutations should be induced accompanied by replacement of Diapr with Arg during the genetic code evolution, meaning that the replacement is probably impossible.

Therefore, it is supposed that the GC code was not the most primitive genetic code, because the inappropriate use of the four amino acids would make it difficult to form water-soluble globular proteins. In addition, the GC code could not evolve to the modern genetic code, because of the difficult exchange from Diapr to Arg.

2.3 Four Column Theory

The four column theory was proposed by Higgs in 2009, based on the facts that four amino acids ([G, A, D, and V]) could be easily formed with prebiotic means and accumulated in large quantities on the primitive earth. The theory assumes that the [GADV]-amino acids were arranged in the four columns as 16 codons encode four respective amino acids, Val: xUx, Ala: xCx, Asp: xAx and Gly: xGx. The small letter "x" means all four bases, U, C, A and G.

Strengths of Four Column Theory The xNx code is considered as the first primitive genetic code encoding four [GADV]-amino acids, based on the analyses of primeval amino acids. The four amino acids are coincident with [GADV]-amino acids encoded by GNC code, which I have proposed. In addition, the four column theory has a characteristic that all base sequences, which were generated by random joining of nucleotides, can be used for [GADV]-protein synthesis.

Weaknesses of Four Column Theory However, the four column theory, in which termination codon was not incorporated, may be problematic about termination of protein synthesis. In the theory, wobble recognition for four bases is assumed at the first and third codon positions. However, the wobble recognition should require a complex mechanism, because two complementary bases usually form only one base pair seeking for the most stable position in the absence of a specific mechanism permitting wobble base pairing. Therefore, it would be impossible to recognize all four bases through wobbling in the most primitive genetic code.

2.4 GCU Code Hypothesis

The idea, that the most primitive genetic code was composed of only one codon (GCU: Ala) coding for Ala, has been also proposed by Trifonov and Bettecken through looking for a common feature in the messenger RNA molecules (Trifonov and Bettecken 1997).

Strengths of GCU Code Hypothesis GCU code hypothesis based on the usage frequency of codon is also certainly consistent with the requirement of the origin of something which should originate from one something.

Weaknesses of GCU Code Hypothesis However, the GCU code hypothesis would be also wrong for the origin of the genetic code. The reason is because poly-L-alanine, which is produced under the GCU code, could not be folded into a water-soluble globular structure but forms nonfunctional α-helix and/or their aggregates as α-helix bundles in water.

2.B. ***The Origin of the Genetic Code-II:*** **How Were the Correspondences Between a Codon and an Amino Acid Established?**

All extant organisms are living under the fundamental life system composed of the six members including gene, genetic code and protein. Above all, it is one of the key issues in life sciences to understand how the first genetic code established and evolved to the modern universal or standard genetic code, because the genetic code, actually tRNA, mediates genetic information written as a codon sequence of a gene to an amino acid sequence of a protein. Therefore, solving the riddle of the origin of the genetic code should lead to understand the ways how gene and protein were formed and even how the first life emerged.

The studies on the origin of the genetic code, exploring how 20 amino acids were arranged in the genetic code table, have been carried out mainly from two viewpoints. One is from the stereochemical theory based on the specific binding between a codon or an anticodon (codon/anticodon) and an amino acid (Shimizu 1982; Yarus 2017), and the other is from the frozen-accident theory, which was proposed by Crick (1968). In the following Sections, it is discussed whether or not the two theories are valid as the origin of the genetic code.

2.5 Stereochemical Theory

I first explain two conditions for that the stereochemical theory holds correct. First one is of course that specific interactions among 20 amino acids and 64, actually 61 codons except 3 termination codons, can be detected. Second one is that the complexes formed between a codon/anticodon and an amino acid effectively function in protein synthesis. Then, I first explain two approaches how researchers have explored to make clear the establishment process of the correspondences between a codon/anticodon and an amino acid for solving the riddle of the origin of the genetic code-II. One is theoretical approach (Shimizu 1982), and the other database analysis of RNA-protein complexes as ribosomes and riboswitches (Yarus 2017).

2.5.1 Search for Specific Interactions Between a Codon/Anticodon and an Amino Acid

It would be natural for many researchers to consider that the establishment process of the genetic code must be attributed to a specific and direct interaction between a codon/anticodon and an amino acid. Therefore, many efforts have been made to solve the problem. For example, Yarus (2017) has reported the results in his recent review article, showing that a significant part of the genetic code were likely established through a direct chemical interaction, based on the precise analyses between a codon/anticodon and an amino acid in RNA-protein complexes (Yarus 2017). Coding triplets, the one codon and four anticodons could be detected for triplet bases in RNAs binding Trp, His, Ile, and Arg. However, the positive results obtained were rather sparse, as only two of 12 Arg triplets (6 codons and 6 anticodons) are significantly implicated by selected sites, or two of six triplets for Ile. In the case of His, only one triplet was found in RNA binding site. He further has proposed that peptides may have been produced directly as depending on an instructive amino acid binding RNA (a DRT; Direct RNA Template) (Yarus 1998, 2017). Although the specific interactions between eight amino acids (Arg, Ile, His, Phe, Trp, Tyr, Gln and Leu) and the corresponding codons/anticodons were certainly detected, inversely stating this, specific interactions between the codons/anticodons and other twelve amino acids (Gly, Ala, Asp, Val, Glu, Pro, Met, Thr, Asn, Lys, Ser, Cys) could not be detected in spite of the strenuous endeavors, meaning that it is difficult or probably impossible to detect the correspondences especially between codons/anticodons and amino acids with a small side chain, which should be easily produced with prebiotic means and should be used in the most primitive genetic code.

2.5.2 Was the Genetic Code Established as Expected by the Stereochemical Theory?

I have confirmed whether the genetic code can be established as supposed by the stereochemical theory. Taking the sizes of amino acid (distance (~3 nm long) between amino residue and carboxyl residue in one amino acid) and three base pairs (distance (~10 nm long) over three base pairs of double-stranded RNA) into consideration, it would be actually impossible to bind one amino acid across three base pairs of codon/anticodon in double-stranded RNA, as seen in three base pairs of AntiC-stem sequences of *Pseudomonas aeruginosa* tRNAs (Ikehara 2019). Therefore, other researchers have searched for specific binding between the codon/anticodon and an amino acid on single-stranded (ss) RNA in the protein-RNA complexes. However, even if the specific binding of an amino acid with three bases (codon/anticodon) of ss-RNA was detected by analyses of the binding sites of protein-RNA complexes as ribosomes and riboswitches, it would be difficult for the triplet base sequence of free ss-RNA to bind with the corresponding free amino acid, because the RNA-protein complexes should be immobilized by many amino acid residues and bases in the protein-RNA complexes and, therefore, binding affinity between three bases in free ss-RNA and a free amino acid should be too small to bind the amino acid with triplet bases.

2.5.3 Can the Complexes Between a Codon/Anticodon and an Amino Acid Function in Protein Synthesis?

There is another serious and critical weakness in the stereochemical theory. That is, even if amino acid could bind with the codon of ss-RNA with a sufficiently strong affinity, the complex could not be utilized to protein synthesis, because it should be quite difficult or probably impossible to form peptide bond between two amino acids among all combinations of the amino acid-triplet base complexes as like protein synthesis in the modern translation system. The reason is because two amino acids immobilized in the two complexes could not locate closely to make peptide bond between the carboxylic group of one amino acid and the amino group of the other amino acid.

In addition, if the stereochemical theory were correct, the protein synthesis under the direct RNA template must be converted to the indirect protein synthesis, which is performed with tRNA in the modern translation system. However it would be also impossible to convert the direct protein synthesis to the indirect protein synthesis.

Strengths of Stereochemical Theory The stereochemical theory on the origin of the genetic code would be also a natural idea, because the theory is considered based on the idea that the genetic code should be established through a stereospecific interaction between a codon or an anticodon and an amino acid. In fact, the specific chemical interactions between several pairs of codons/anticodons and amino acids were confirmed with RNA-protein complexes, as ribosomes and riboswithes.

Weaknesses of Stereochemical Theory However, it would be difficult or probably impossible to detect the whole correspondences especially between a codon/anticodon and an amino acid with a small side chain, which should be used in the most primitive genetic code. Furthermore, peptide bond formation between two amino acids through amino acid-RNA complexes would be also impossible. As the results, Kun and Radvanyi (2018) has described in their paper that the origin of the genetic code is a "notoriously difficult problem". Thus, it must state that the reason, why 20 kind of amino acids are so beautifully arranged in the genetic code table, is unfortunately unknown still now (Fig. 7.3). In addition, it would be impossible to explain

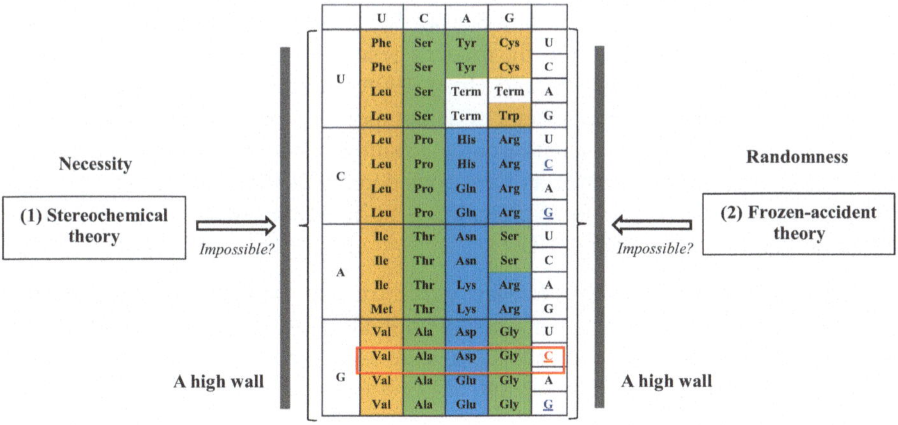

Fig. 7.3 Universal (standard) genetic code table. From deciphering the genetic code, it was found that the genetic code has been so beautifully constructed, for example, as arranging hydrophobic amino acids at the most left column. It would be difficult to explain the establishment processes with "Stereochemical theory" or "Frozen-accident theory" as explained in the text. Instead, the reason, why the genetic code is formed so beautifully, remains unsolved (this figure is essentially the same with Fig. 7.1). GNC primeval code is shown with a red square

the way how direct correspondences between a codon/anticodon and an amino acid as suggested by the stereochemical theory could be transferred to the indirect correspondences between an anticodon and an amino acid, which are realized with tRNA and ARS. These also indicate that the stereochemical theory is something wrong. Therefore, the origin of the genetic code could not be solved by the stereochemical theory.

2.6 Crick's Frozen-Accident Theory

Nearly 50 years ago, Crick (1968) tackled the problem how the universal genetic code has become in the present form. However, the problem, about how correspondences between a codon/anticodon and an amino acid were determined, perplexed Crick, because he had to consider that the establishment process might not be explained from the standpoint of the stereochemical theory, based on the results obtained until that time. Then, he proposed the "frozen-accident theory" on the origin of the genetic code, assuming that the genetic code was formed during evolution of the genetic code by the "frozen-accident" method.

However, it would be impossible to randomly construct the genetic code having the systematized features as robustness against base substitutions, especially at the first and the third codon positions (Fig. 7.3). Therefore, there are many criticisms in the "frozen-accident theory". For example, Koonin has even described in his paper that the frozen accident can be considered as the default theory of code evolution because it does not imply any specific interactions between the codon/anticodon and an amino acid (Koonin and Novozhilov 2009; Koonin 2017). Therefore, it is even considered at the present time that it would be impossible to accidentally form the whole genetic code during evolution. The idea is supported by calculation results that the appearance probability of the modern genetic code through random process is quite low as $\sim 10^{-6}$ (Freeland and Hurst 1998; Koonin and Novozhilov 2009; Koonin 2017). The universal genetic code composed of 64 codons and 20 amino acids is too complex to establish it through the simply random process at one stroke. Thus, nowadays, many researchers have considered that the establishment process of the genetic code cannot be explained by the frozen-accident theory, which is based on the randomness (Fig. 7.3).

Strengths of Frozen-Accident Theory Crick tried to explain, the way how the amazing correspondences between a codon/anticodon and an amino acid, which can be seen in the universal genetic code table, were established, because it would intuitively seem to him that the correspondences could not be explained by the stereochemical theory.

Weaknesses of Frozen-Accident Theory As can be easily understood from the smallness of the appearance probability of the modern genetic code through random process, it would be obvious that the "frozen-accident theory" would be wrong for the origin of the genetic code (Freeland and Hurst 1998; Koonin and Novozhilov 2009; Koonin 2017).

3 My Idea on the Origin of the Genetic Code

3.1 For What Purpose Was the First Genetic Code Created?

Genetic code is a table showing correspondences between a codon and an amino acid. Such genetic code could not be formed among the six members of the fundamental life system at first, because the genetic code would be meaningless in the absence of gene, protein and others. Therefore, it would be obvious that the genetic code was formed and written into the table as results that the correspondences were determined upon formation of tRNAs directly connecting amino acid with codon actually anticodon. This means that the genetic code was simply formed as a result of the origin and evolution of tRNA. Then, it is discussed in the following Sections, how the genetic code originated from what code and has become to the present state.

3.A. ***The Origin of the Genetic Code-I:*** **From What Code the Genetic Code Originated?**
Whenever the origin of something, such as the genetic code, is investigated, there is a serious problem, as that it is almost impossible to make clear completely, what events took place in the past, especially on the primitive Earth about 3.8 ~ 4.0 billion years ago. However, I would like to emphasize that the problems of the origins of gene, the genetic code and protein are too important to avoid pursuing them, because the origins are always relevant to the fundamental properties of extant gene, the genetic code and protein. I have proposed GNC-SNS primitive genetic code hypothesis on the origin and evolution of the genetic code (Ikehara et al. 2002). I will explain effectiveness of the hypothesis in detail in the following Section.

3.2 GNC-SNS Primitive Genetic Code Hypothesis

My investigation of the origin of the genetic code started from studying on the mechanism, under which entirely new (EntNew) genes have been formed in presently existing microorganisms, independent of the origin of the genetic code. That, fortunately, led me to the research on the origin of the genetic code. For that purpose, data of microbial water-soluble globular proteins and their genes obtained from GenomeNet Database (GenBank Overview (nih.gov)) were analyzed using the six conditions (hydropathy, α-helix, β-sheet and turn/coil structure formabilities, acidic amino acid, and basic amino acid compositions) for water-soluble globular protein formation. The flow of my investigation is described in brief below, as I described it in detail in my book, which was published from LAP LAMBERT Academic Publishing (Ikehara 2016).

3.2.1 SNS Primitive Genetic Code Hypothesis

First, we investigated on a field for creation of an EntNew gene by using the six conditions for formation of water-soluble globular protein in order to solve the origin of contemporary EntNew gene, which should be created in extant organisms even nowadays. From the results, the GC-NSF(a) hypothesis on the origin of contemporary gene was obtained, suggesting that EntNew genes originated from non-stop frames on antisense sequence of GC-rich genes (GC-NSF(a)s) (Ikehara and Okazawa 1993; Ikehara et al. 1996).

After the proposition of the GC-NSF(a) hypothesis, I noticed that base composition format at three codon positions of both highly GC-rich genes (65 ~ 70%) and their GC-NSF(a) sequences are roughly SNS or [(G/C)N(C/G)] (Fig. 7.4) (Ikehara and Yoshida 1996; Ikehara 2002; Ikehara et al. 2002).

Then, I considered a possibility that SNS code encoding only ten amino acids ([GADV]-amino acids plus Glu [E], Leu [L], Pro [P], His [H], Gln [Q], and Arg [R]) was a primitive genetic code. To confirm whether or not the SNS code has a coding ability for formation of a water-soluble globular protein, random numbers were generated in a frame of SNS at the three codon positions in a computer. Dot was plotted, if an imaginary protein encoded by the SNS code satisfied the six structural conditions for formation of a water-soluble globular protein (Fig. 7.5) (Ikehara, K. 2002; Ikehara et al. 2002).

From the results, it was found that the computer-generated SNS code encoding 10 amino acids satisfies the six conditions, when the compositions of G and C at the first codon position were at around 55% and 45%, respectively, and when every base was contained roughly at a ratio of about one-fourth each at the second codon position (Fig. 7.5). Base compositions at the third position could not be restricted into a small range due to the degeneracy of the genetic code at that position. However, this also means that polypeptide chain composed of SNS-encoding ten amino acids should be folded into water-soluble globular structure at a high probability, because the SNS code satisfies the six conditions for formation of a water-soluble globular structure.

3.2.2 GNC Primeval Genetic Code Hypothesis

Next, I supposed that the SNS code must originate from a simpler code. To search for a genetic code more ancient or simpler than the SNS code, we used the four protein structure indexes (hydropathy, α-helix, β-sheet, and turn/coil formabilities) as minimum conditions for formation of a water-soluble globular protein with appropriate secondary and tertiary structures. From the results, it was found that [GADV]-protein encoded by GNC code satisfies well the four conditions, when about equal amounts of [GADV]-amino acids are contained in a protein (Fig. 7.6) (Ikehara et al. 2002).

The four [GADV]-amino acids are excellent for formation of the respective secondary structures and an active center on [GADV]-protein, as [G] is an amino acid, which is effective for turn or coil formation, and [A] and [V] are effective for α-helix

Fig. 7.4 Average base compositions at three codon positions of seven GC-rich *P. aeruginosa* genes. Open and closed bars show the average base compositions of seven *P. aeruginosa* GC-rich genes and the corresponding GC-NSF(a)s, respectively

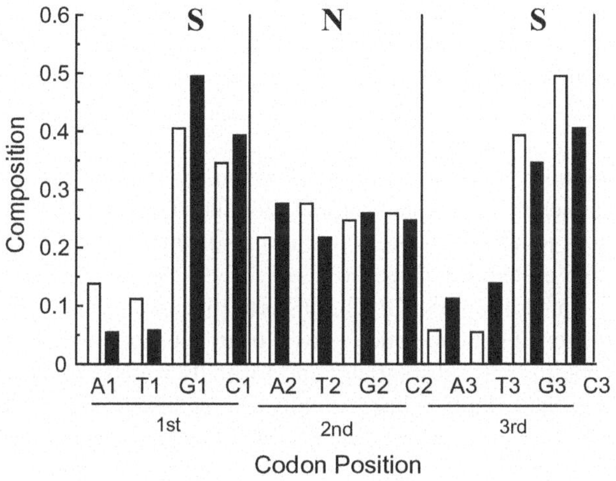

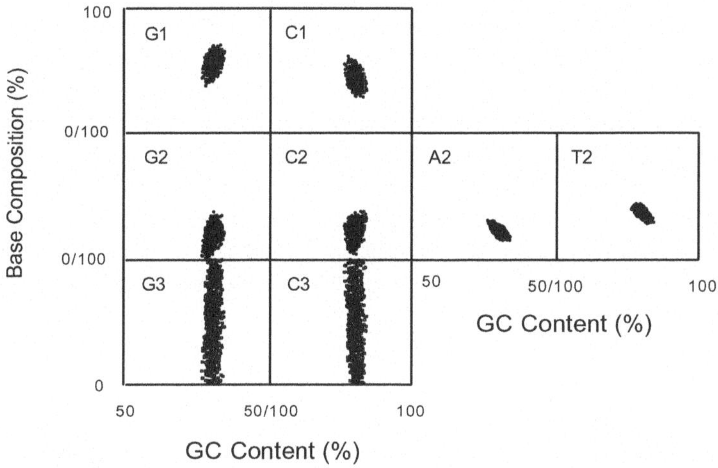

Fig. 7.5 Dot representation of base compositions at three base positions in codon, which were selected by determining whether imaginary protein computer-generated under the SNS coding system satisfies the six conditions for water-soluble globular structure formation

and for β-sheet formations, respectively (Berg et al. 2002). Both hydrophobic ([V]) and hydrophilic ([D]) amino acids are also fortunately included in the four amino acids encoded by the GNC code. This helps the polypeptide chain to fold into a stable globular structure in water. Furthermore, the combination of the four amino acids is the simplest one out of those of the four amino acids selected from the 20 natural amino acids, suggesting that the GNC code could be the most ancient genetic code. This conclusion is the same as antiquity of the "GNC" code, which was first proposed by Eigen and Schuster (1977, 1979) from an independent standpoint of coding ability of protein. It is also well known from Miller's type electric discharge experiments that the four [GADV]-amino acids could be easily formed and accumulated in large quantities on primitive Earth (Miller and Orgel 1974,

Van der Gulik et al. 2009, Higgs 2009). Furthermore, the idea that the universal genetic code originated from GNC code encoding four [GADV]-amino acids is also supported by the fact that [GADV]-amino acids are consistent with the chronological order of amino acids, which was published by Trifonov (2000).

These support the idea, that the amino acid composition containing [GADV]-amino acids at roughly equal amounts, is a protein 0th-order structure, because it suggests that a water-soluble globular [GADV]-protein could be produced by random joining of [GADV]-amino acids encoded by the first GNC genetic code at a high probability (Fig. 7.6).

In addition, the GNC-SNS primitive genetic code hypothesis represents that the universal genetic code evolved from substantially singlet but formally triplet GNC code to both substantially and formally triplet universal genetic code through substantially doublet but formally triplet SNS code. In this way, we have proposed GNC-SNS primitive genetic code hypothesis, suggesting that the universal or standard genetic code originated from GNC code encoding four [GADV]-amino acids, through SNS code (Fig. 7.7) (Ikehara 2002; Ikehara et al. 2002).

Strengths of GNC-SNS Primitive Genetic Code Hypothesis Both formation process of the most primitive GNC code through random processes and evolutionary process from the GNC code to modern universal genetic code can be well explained. The GNC-SNS primitive genetic code hypothesis has been proposed based on the ability of water-soluble globular protein formation.

Weaknesses of GNC-SNS Primitive Genetic Code Hypothesis The greatest weakness of the GNC-SNS primitive genetic code hypothesis is that it is not confirmed by experiments whether or not a [GADV]-polypeptide chain encoded by GNC code

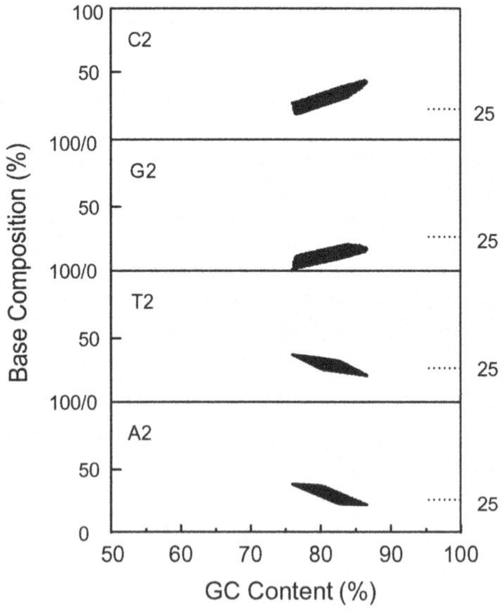

Fig. 7.6 Dot representation of base compositions at the second base position in the codon, which were selected by determining whether imaginary protein computer-generated under the GNC coding system satisfies the four structural conditions for water-soluble globular structure formation (note that this Fig. 7.6 is the same with Chap. 3: Fig. 3.10)

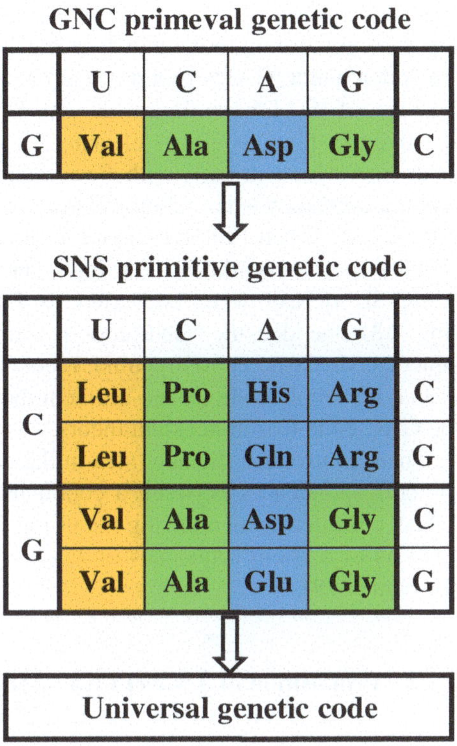

Fig. 7.7 GNC-SNS primitive genetic code hypothesis assuming that the universal or standard genetic code (both formally and substantially triplet code) originated from GNC primeval genetic code (formally triplet but substantially singlet code) through SNS primitive genetic code (formally triplet but substantially doublet code) composed of 10 amino acids encoded by sixteen codons. Brown and blue color boxes indicate hydrophobic and hydrophilic amino acids, respectively

and a polypeptide chain composed of ten amino acids encoded by SNS code can be folded into water-soluble globular structure at a high probability as suggested by the hypothesis.

3.B. ***The Origin of the Genetic Code-II:*** **How Were the Correspondences Between a Codon/Anticodon and an Amino Acid Established?**
During I considered the process how four non-specific primitive AntiC-SL tRNAs were converted to four specific [GADV]-AntiC-SL tRNAs (Chap. 6, Fig. 6.9), I suddenly hit upon a new theory on the origin of the genetic code by accident. That is the GNC code frozen-accident theory. I will explain the theory in the following Section.

3.3 GNC Code Frozen-Accident Theory

3.3.1 A Novel Idea for Solving the Enigma of the Origin of the Genetic Code-II

I have proposed a novel hypothesis on the origin of tRNA in this book, assuming that modern tRNAs originated from one nonspecific AntiC-SL tRNA, with which one of four [GADV]-amino acids was randomly bound, and evolved later into four specific AntiC-SL tRNAs through the four nonspecific AntiC-SL tRNAs (Chap. 6, Fig. 6.9). According to the new hypothesis, the correspondences between a codon/anticodon and a [GADV]-amino acid must arise at one time. However, the specific bindings of [GADV]-amino acids with GNC codons have not been detected irrespective of strenuous efforts of many researchers, suggesting that [GADV]-amino acids cannot bind strongly with the corresponding codons or anticodons. Then, how was the correspondence relations between [GADV]-amino acids and GNC codons established? When I thought about a role of the most ancient AntiC-SL tRNAs in [GADV]-protein synthesis or during I considered the process how four non-specific AntiC-SL tRNAs were converted to four specific [GADV]-AntiC-SL tRNAs (Fig. 7.8), one idea, the "GNC code-frozen accident theory", about the origin of the genetic code flashed into my mind. I feel now the flash of Crick about 50 years ago, as that his whole code frozen-accident theory might be unconsciously inspired by the GNC-code frozen-accident theory.

Then, I introduce the new idea in the next Section, explaining the process how the correspondences between a codon/anticodon and an amino acid were established, considering as referring the origin of tRNA, which I recently obtained from analysis of 5′-stem sequences of *P. aeruginosa* PAO1 tRNAs (Chap. 6: Sect. 4) (Ikehara 2019).

3.3.2 Proposition of a Novel Idea on the Origin of the Genetic Code or "GNC Code-Frozen Accident Theory"

The GNC code-frozen accident theory assumes that the correspondences between a GNC codon and a [GADV]-amino acid had been accidentally and randomly established and that modern universal genetic code was established by piling up of amino acids, which have abilities necessary to improve protein function produced under the previous genetic code, on the basis of the primeval GNC code one by one.

(1) Necessary usage of [GADV]-amino acids and GNC codons

1. Use of the four [GADV]-amino acids in the first primeval genetic code was inevitable: Needless to state, the genetic code is constructed with the combination of a codon and an amino acid. As already explained in detail in Chaps. 3 and 6, it has been confirmed that amino acids used in the most primitive protein were [GADV]-amino acids, because the four amino acids are the simplest combination of four amino acids, of which by random joining, water-soluble globular proteins could be produced, and, therefore, are the most suitable to the first genetic code. Furthermore, [GADV]-amino acids should be easily produced with prebiotic means and accumulated on the primitive Earth. Therefore, [GADV]-amino acids were necessarily used for synthesis of the most primeval proteins.

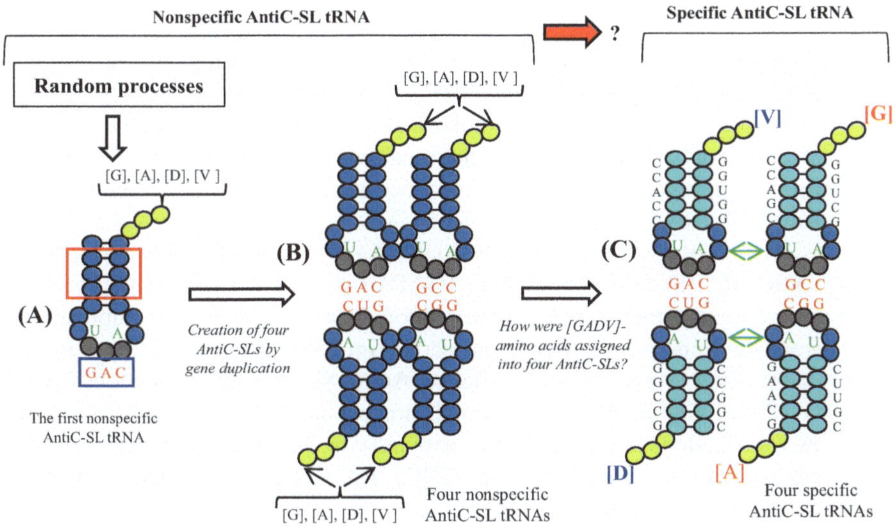

Fig. 7.8 The "anticodon-stem loop hypothesis" on the origin of tRNA. (**a**) The hypothesis assumes that modern tRNAs originated from one nonspecific primeval AntiC-SL tRNA carrying one of four [GADV]-amino acids selected randomly. The nonspecific AntiC-SL had equipped with CCA-end shown with three small yellow circles at the 3′-end. (**b**) Successively, four nonspecific AntiC-SLs were created by gene duplication of the first AntiC-SL gene. (**c**) After that, two pairs of [GADV]-amino acids (Gly/Ala and Asp/Val) were randomly assigned to two AntiC-SL pairs, one is AntiC-SL (GCC) and AntiC-SL (GGC), and the other is AntiC-SL (GAC) and AntiC-SL (GUC), and were frozen. The anticodons of two AntiC-SLs were necessarily selected because of a high stability of triplet base pairs (GCC/GGC and GAC/GUC) (Taghavi et al. 2017). Consequently, the four specific primeval AntiC-SL tRNAs carrying one specific amino acid (shown by blue and red letters in bracket) were obtained. The specific tRNA is characterized by 5′ accepter stem sequence (light cyan circles) and anticodon (dark gray circles). Green double-headed arrows indicate strong interaction between two AntiC-loops through A-U base pair. Small grayish blue circles mean nucleotides in nonspecific AntiC-stems

2. Use of the four GNC codons in the first primeval genetic code was also inevitable: The reasons, why AntiC-SL composed of 17 nucleotides became the first primeval tRNA, are not only because the AntiC-SL is small and sufficiently stable but also because two stable base pairings, GCC/GGC and GUC/GAC, could be formed between two vertically aligned AntiC-SLs carrying the anticodons to make efficiently peptide bond between two amino acids bound with two AntiC-SLs juxtaposed side by side (Fig. 7.8). Therefore, the four GNCs were necessarily used for the first anticodons or codons.

(2) Formation of nonspecific AntiC-SL tRNAs.

1. Formation of one nonspecific AntiC-SL tRNA: One small but sufficiently stable AntiC-SL was produced through a random process and repeated error-prone polymerization of nucleotides and degradation of less stable oligonucleotides (Chap. 6: Fig. 6.9). One meaningful RNA structure, which was formed through the random process, became the first primeval nonspecific AntiC-SL tRNA (Fig. 7.8a).
2. Formation of four nonspecific [GADV]-AntiC-SL tRNAs: Four nonspecific AntiC-SL tRNAs for producing [GADV]-protein with random [GADV]-amino acid sequence were created through two times of gene duplications from the gene for the first nonspecific AntiC-SL (Fig. 7.8b). Two dimers composed of two vertically aligned AntiC-SLs having anticodons GCC/GGC and GUC/GAC were formed through stable base-pairs between the two anrticodons of the AntiC-SLs to synthesize [GADV]-peptides more efficiently than before. The first ss-$(GNC)_n$ RNA could be created by random joining of anticodons of the four nonspecific AntiC-SLs for the first time.

(3) Formation of four specific [GADV]-AntiC-SL tRNAs and establishment of the GNC primeval genetic code.

Formation of four specific [GADV]-AntiC-SL tRNAs: After formation of the four nonspecific [GADV]-AntiC-SL tRNAs, two pairs of Gly/Ala and Asp/Val were accidentally or randomly assigned to two pairs of AntiC-SL tRNAs carrying one anticodon of GCC/GGC and GUC/GAC, respectively (Fig. 7.8c). The affair generating four specific [GADV]-AntiC-SL tRNAs from four nonspecific AntiC-SL tRNAs became the critical point of GNC code-frozen accident theory (Fig. 7.9a). In other words, anticodons for Gly/Ala and Asp/Val were randomly assigned in the frames of two AntiC-SL pairs carrying GCC/GGC and GUC/GAC anticodon, independently of stereochemical interaction, to make correspondences between [GADV]-mino acids (Gly, Ala, Asp and Val) and GNC codons (GGC, GCC, GAC and GUC) as seen in the modern genetic code, respectively (Figs. 7.7 and 7.9a). The random assignment of four [GADV]-amino acids to the respective specific AntiC-SL tRNAs carrying GNC anticodon made it possible to lead to formation of the specific [GADV]-ARSs having ordered [GADV]-amino acid sequence for the first time. Thus, the GNC primeval genetic was established through formation of the four specific AntiC-SL tRNAs (Chap. 6: Fig. 6.9).

(4) Evolutionary process from GNC code to SNS code

Thereafter, one amino acid satisfying the following conditions for synthesis of protein with a higher function, was selected and added into the GNC primeval genetic code encoding [GADV]-amino acids one by one, as was expected by the *[1)] coevolution theory (Wong 1975; Di Giulio 2008) and the *[2)] adaptive theory (Knight et al. 2007) (Fig. 7.9).

*1) Coevolution theory is an idea assuming that amino acids were captured into the previously existed genetic code in order of amino acids newly accumulated in cells upon development of amino acid metabolism and, therefore, the usage of the amino acids in the genetic code and amino acid metabolism progressed in parallel. Therefore, it is expected that amino acid newly synthesized in the amino acid metabolism should be captured into the genetic code used previously. However, correspondence between a codon and an amino acid can not be determined based on only the coevolution theory, because it can not be determined which codon must be used by the amino acid newly synthesized (refer to the description of coevolution in Chap. 5, Sect. 3.5).

*2) Adaptive theory could solve the cumbersome problem. In other words, the evolutionary process of the genetic code can be explained by using both the coevolution theory and the adaptive theory complementarily. That is, it is considered that the universal genetic code has been established by embedding newly synthesized amino acids at appropriate positions into the previously prepared genetic code tables one by one according to the adaptive theory. In a review article of Moura et al. (2010), it is described that the adaptive theory postulates that the evolution of the genetic code is mainly driven by the selective forces that minimize the effects of protein synthesis errors, being them from mutational origin or from mRNA misreading. The observation that amino acids with similar chemical properties are assigned to similar codons plus statistical and computational evidence for a strong bias towards error minimization pressure in the code provide important support for this theory.

1. Addition of Glu to the primeval GNC code: The incorporation of amino acid into the previously existed genetic code was dependent on development of amino acid metabolic pathways as deduced by the coevolution theory (Wong 1975; Wong et al. 2016; Di Giulio 2008). In addition, an amino acid, which can compensate for an insufficient structure formability of proteins produced under the genetic code used one step before, was incorporated into the genetic code to minimize the effects of protein synthesis errors as was expected by the adaptive theory (Knight et al. 2007). For example, α-helix forming acidic amino acid, Glu, was captured into GNC code to compensate the deficient ability in the [GADV]-amino acids (Fig. 7.9b). Although incorporation of Glu into the GNC code could compensate for the insufficient α-helix forming ability of [GADV]-amino acids, the incorporation of the hydrophilic Glu into the GNC code would also cause an excess hydrophilicity of [GADV+E]-protein because of duplicate use of the hydrophilic acidic amino acids, Asp and Glu. That would be the reason why both Asp and Glu in the modern genetic code use two-codon boxes in stead of four codon box used by three other [GAV]-amino acids. Degeneracy of the genetic code started from the duplicate codon usage of Gly, Ala and Val in GNS code to correct the ratio of [GADEV] = 1:1:1:1:1 to 2:2:(1:1):2. Thus, the degeneracy of the genetic code was established at a level. Consequently, the possibility of creation of gene/protein with a higher function became large (Fig. 7.9a-(2–5)).

Thus, a new AntiC-SL, which was formed under the conditions for selection of codon/anticodon and the stem sequence of AntiC-SL, was assigned to the newly selected amino acid. The newly established genetic code was frozen at every evolutionary stage succeeding to the first GNC code, because one codon change of an amino acid in the genetic code, which was once established, should induce multiple amino acid replacements. AntiC-SL carrying a new anticodon, which was formed by one base substitution, was generally used for translating a new codon created by

(A)

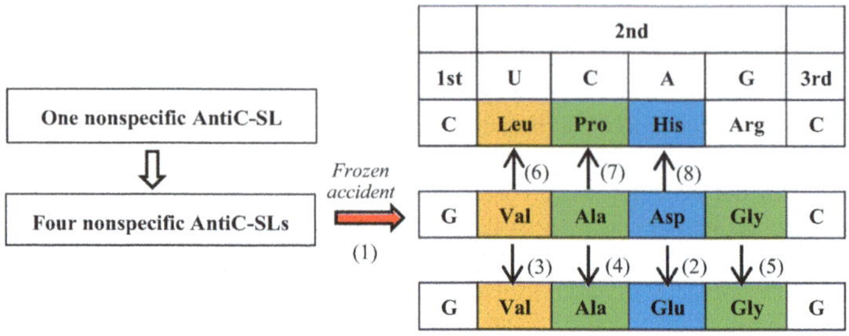

(B)

G-strat codons

	HP	α-Helix	β-Sheet	Turn
Gly	1	0.56	0.92	1.64
Ala	1.6	1.29	0.9	0.78
Asp	-9.2	1.04	0.72	1.41
Val	2.6	0.91	1.49	0.47

Glu	-8.2	1.44	0.75	1

C-strat codons

	HP	α-Helix	β-Sheet	Turn
Leu	2.8	1.3	1.02	0.59
Pro	-0.2	0.52	0.64	1.91
His	-3	1.22	1.08	0.69
Gln	-4.1	1.27	0.8	0.97

Fig. 7.9 (**a**) Evolutionary process of the genetic code. (1) The GNC code encoding [GADV]-amino acids was established by random assignment of Gly and Ala to an AntiC-SL pair carrying anticodon, GCC or GGC, and Asp and Val to AntiC-SLs carrying anticodon, GUC or GAC, respectively (Fig. 7.8). (2) After the establishment of the GNC code, Glu was incorporated into the GNC code. (3), (4) and (5) Synonymous codons of Val, Ala and Gly were captured to complement deficient abilities for secondary and tertiary structure formation caused upon the incorporation of hydrophilic Glu into GNC code. (6), (7) and (8) Three amino acids were further incorporated into the GNS code, in order of Leu, Pro and His. Thus, the GNC code, which was established by random assignments, evolved to SNS code and to the universal genetic code, as incorporating necessary amino acids into the previous genetic code, one by one. The order of amino acid incorporation into the GNC code is consistent with the origin and evolutionary process of tRNA (Chap. 6: Fig. 6.12) (Ikehara 2019) and GNC-SNS genetic code hypothesis (Fig. 7.7) (Ikehara et al. 2002). (**b**) Protein structure formabilities of amino acids. Up-left and down-left tables indicate protein structure formabilities of [GADV]-amino acids and Glu, respectively. Up-right table is those of Lue, Pro, His and Gln. HP means hydrophobicity index (Berg et al. 2002). Darker color indicates the respective higher protein structure formabilities

one base replacement on a previous genetic sequence or produced in a new gene (Ikehara 2019). Therefore, a triplet base sequence obtained by one base replacement from a previously existed codon/anticodon was used as a new codon/anticodon for the amino acid, which was newly incorporated into the code one step before.

2. Capture of Leu, Pro, His and Gln to form GNS code: Leu is a highly hydrophobic α-helix forming amino acid (Fig. 7.9b). Therefore, Leu would be incorporated into the GNS genetic code encoding [GADV+E]-amino acids in order to strengthen insufficient hydrophobicity and α-helix forming ability of the GNS

code (Fig. 7.9a-(6)). Successively, Pro, which is an exclusively turn/coil forming amino acid, was captured into the previous code in order to compensate for insufficient turn/coil forming ability of [GADV+EL]-amino acids (Fig. (A)-(7) and 6-7 (B)). Basic amino acid, His, which is a comparatively high α-helix and β-sheet forming amino acid, would be incorporated into the code to compensate for insufficient abilities, especially to add a positive charge deficient to [GADV+ELP]-protein (Figs. 7.9 (A)-(8) and (B)). Finally, Gln and Arg would be captured into the previous genetic code to create SNS code in turn, as similarly reinforcing insufficient abilities in structure and function of protein.

(5) Evolutionary process from SNS code to the universal genetic code.

1. Establishment of the universal genetic code: After the completion of SNS primitive genetic code, it became possible to recognize two codons SNY or SNR, by one AntiC-SL tRNA through wobble base pairing to form SNN code composed of two or four codon boxes. Thus, the region of degeneracy of the genetic code could be enlarged. Successively, amino acids encoded by A-start codons were incorporated into the SNN code, one by one. Furthermore, amino acids encoded by U-start codons were captured to complete the beautiful universal genetic code. Three codons, UAA, UAG and UGA, remained unused to efficiently induce termination of protein synthesis.
2. Formation of tRNA with anticodon facing outward from anticodon-loop: Triplet anticodon bases carried by AntiC-SL must be turned outward from the anticodon loop to translate a genetic sequence to an amino acid sequence efficiently. Four AntiC-SLs carrying GNC anticodon could be accomplished without any chemical modification to the bases of anticodon loop (tRNAdb (Universitat Leipzig)). However, AntiC-SLs carrying anticodon other than GNCs must be chemically modified to turn outward the anticodon to read codons on the genetic sequence. Therefore, AntiC-SLs or tRNAs, of which anticodon can be turned outward with a simpler chemical modification, was used at earlier evolutionary stage. Thus, the order of usable codons was determined by a kind of modified base neighboring the anticodon. At that time, codons in another row were used, after all N-start codons were exhausted in order of G-start, C-start and A-start codons.

Thus, amino acids were selected and incorporated into the previous genetic code, one by one, as piling up on the basis of GNC primeval genetic code. The genetic code was frozen at every evolutionary stage, because any change of the genetic code caused multiple mutations leading to lethal effects to the organisms.

Strengths of GNC Code-Frozen Accident Theory It could be concluded according to the necessary conditions for the origin of something that GNC code-frozen accident theory is valid as the origin of the genetic code, because both formation process of the most primitive GNC code through random process and evolutionary process from the GNC code to modern universal genetic code can be reasonably explained as follows.

The first GNC code was frozen accident into four AntiC-SL tRNAs carrying one of four GNC codons as an anticodon. Thereafter, the genetic code evolved by capturing a new amino acid-tRNA pair as piling up onto the GNC code, as expected by the coevolution theory and the adaptive theory on the origin and evolution of the genetic code.

Weaknesses of GNC Code-Frozen Accident Theory Thus, I have considered that the most primitive GNC code was established frozen accidentally and evolved to the universal genetic code through SNS primitive genetic code, as supposed by the GNC code-frozen accident theory and GNC-SNS primitive genetic code hypothesis. However, simultaneously, I have considered that there is a great weakness in my idea on the origin of the genetic code that several matters assumed in the theory have not been confirmed by experiments as described below.

1. Was the first GNC genetic code really established frozen accidentally among four [GADV]-amino acids and four AntiC-SL tRNAs as assumed by the the GNC code-frozen accident theory?
2. Was the universal genetic code really established by piling up amino acids one by one on the basis of the GNC primeval genetic code, as assumed by the GNC code-frozen accident theory, and so on.

4 Discussion

As a matter of course, considering the origin of the genetic code is to explore from what code the modern genetic code originated. There are two different sides for exploring the origin of the genetic code as follows.

1. From what the genetic code originated?
2. How were the correspondences between a codon and an amino acid established?

Furthermore, something assumed as the origin of the genetic code-I and -II must satisfy two following conditions to be valid. The first one is that the something assumed as the origin must be produced through random process and the second one is that the something must evolve to modern genetic code as described above.

As seen in Fig. 7.10, the GNC code was established by random process and was accidentally frozen as suggested by the GNC code-frozen accident theory. The reason why the GNC code was used for the first genetic code is because the smallest and stablest AntiC-SL carrying GNC anticodon, which was produced during repeated synthesis and degradation of oligonucleotide, was simply and fortunately adopted as the first tRNA. In addition, it is considered that the universal genetic code originated from the GNC code through SNS primitive genetic code (Figs. 7.7 and 7.10). Therefore, modern organisms use the 20 amino acids as a result that the formally triplet but substantially singlet GNC code was used as the first genetic code. Consequently, modern organisms are prospering on this planet.

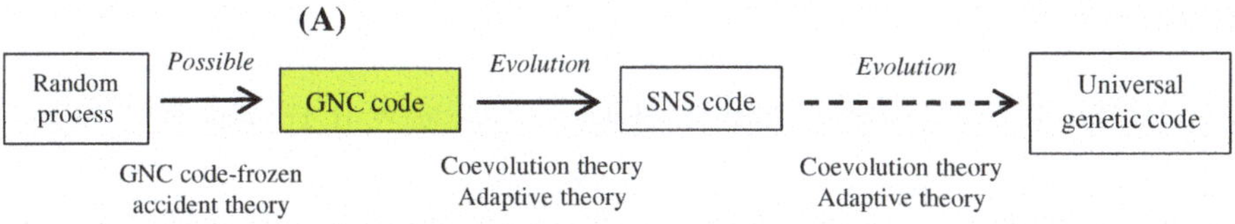

Fig. 7.10 The origin and evolutionary process of the universal genetic code can be well explained by GNC code-frozen accident theory and GNC-SNS primitive genetic code hypothesis without any large contradiction. (**a**) First, GNC code was frozen-accidentally established using four [GADV]-amino acids and four AntiC-SL tRNAs, both of which were produced with the respective random processes. The first genetic code evolved to the universal genetic code through SNS code as expected by the coevolution theory and the adaptive theory on the origin and evolution of the genetic code. It would be probably impossible to explain the origin and evolutionary processes with hypotheses proposed by other researchers. See also Sect. 2 in the text

Use of the 20 amino acids for protein synthesis would be sufficient for extant organisms to inhabit on this planet. This can be easily understood from the use of 6 codons by three amino acids, Leu, Arg and Ser, and from degeneracy at the third codon position. It is also considered that a high ability of [GADV]-amino acids encoded by GNC code makes it possible to produce the present high level of proteins exhibiting various amazing functions.

5 Conclusion

The genetic code is a table representing correspondences between a codon and an amino acid, which are mediated by tRNA. Therefore, the essence of the genetic code can be described as follows.

1. Both the origin and the evolutionary process of the universal genetic code are determined by tRNA. For example, the reason, why the first codons and the first amino acids were GNC codons and [GADV]-amino acids, respectively, are simply because the first four AntiC-SL tRNAs used the four GNC anticodons and the four [GADV]-amino acids for immature [GADV]-protein synthesis, although the correspondences were accidentally frozen.
2. The use of GNC code by the AntiC-SL tRNAs determined the frame of the universal genetic code, which are using triplet code. The modern genetic code was established as a result of the subsequent evolution of tRNA, which is consistent with the coevolution theory and the adaptive theory.

Thus, the essence of the genetic code is an expression of the results, which are determined by the origin and evolution of tRNA.

References

Berg JM, Tymoczko JL, Stryer L (2002) Biochemistry, 5th edn. W. H. Freeman and Company, New York

Crick FHC (1968) The origin of the genetic code. J Mol Biol 38:367–379

Di Giulio M (2008) An extension of the coevolution theory of the origin of the genetic code. Biol Direct 3:37

Eigen M, Schuster P (1977) The hypercycle. A principle of natural self organization. Part A: Emergence of the hypercycle. Naturwissenschaften 64:541–565

Eigen M, Schuster P (1979) Hypercycle: a principle of natural self-organization. Springer, Heidelberg

Freeland SJ, Hurst LD (1998) The genetic code is one in a million. J Mol Evol 47:238–248

Hartman H, Smith TF (2019) Origin of the genetic code is found at the transition between a thioester world of peptides and the phosphoester world of polynucleotides. Life 9:69

Higgs PG (2009) A four-column theory for the origin of the genetic code: tracing the evolutionary pathways that gave rise to an optimized code. Biol Direct 24:4–16

Ikehara K (2002) Origins of gene, genetic code, protein and life: comprehensive view of life system from a GNC-SNS primitive genetic code hypothesis. J Biosci 27:165–186

Ikehara K (2016) GADV hypothesis on the origin of life – life emerged in this way!? LAP LAMBERT Academic Publishing, Saarbrucken

Ikehara K (2019) The origin of tRNA deduced from *Pseudomonas aeruginosa 5′* anticodon-stem sequence: anticodon stemloop hypothesis. Orig Life Evol Biosph 49:61–75

Ikehara K, Okazawa E (1993) Unusually biased nucleotide sequences on sense strands of Flavobacterium sp. genes produce nonstop frames on the corresponding antisense strands. Nucleic Acids Res 21:2193–2199

Ikehara K, Yoshida S (1996) SNS hypothesis on the origin of the genetic code. Viva Origino 26:301–310

Ikehara K, Amada F, Yoshida S et al (1996) A possible origin of newly-born bacterial genes: significance of GC-rich nonstop frame on antisense strand. Nucleic Acids Res 24:4249–4255

Ikehara K, Omori Y, Arai R et al (2002) A novel theory on the origin of the genetic code: a GNC-SNS hypothesis. J Mol Evol 54:530–538

Knight RD, Freeland SJ, Landweber LF (2007) Adaptive evolution of the genetic code. In: de Pouplana LR (ed) The genetic code and the origin of life. Springer, Boston, MA, USA pp 201–220

Koonin EV (2017) Frozen accident pushing 50: stereochemistry, expansion, and chance in the evolution of the genetic code. Life (Basel) 7(2):22

Koonin EV, Novozhilov AS (2009) Origin and evolution of the genetic code: the universal enigma. IUBMB Life 61:99–111

Kun Á, Radvanyi Á (2018) The evolution of the genetic code: impasses and challenges. Biosystems 164:217–225

Miller SL, Orgel LE (1974) The origins of life on the earth. Prentice-Hall, Englewood Cliffs

Moura GR, Paredes JA, Santos MAS (2010) Development of the genetic code: insights from a fungal codon reassignment. FEBS Lett 584:334–341

Nirenberg M (1968) Concluding remarks of his nobel lecture

Shimizu M (1982) Molecular basis for the genetic code. J Mol Evol 18:297–303

Taghavi A, van der Schoot P, Berryman JT (2017) DNA partitions into triplets under tension in the presence of organic cations, with sequence evolutionary age predicting the stability of the triplet phase. Q Rev Biophys e15. https://doi.org/10.1017/S0033583517000130

Trifonov EN (2000) Consensus temporal order of amino acids and evolution of the triplet code. Gene 261:139–151

Trifonov EN, Bettecken T (1997) Sequence fossils, triplet expansion, and reconstruction of earliest codons. Gene 205:1–6

Van der Gulik P, Massar S, Gilis D et al (2009) The first peptides: the evolutionary transition between prebiotic amino acids and early proteins. J Theor Biol 261:531–539
Watson JD, Hopkins NH, Roberts JW et al (1987) Molecular biology of the gene, 4th edn. Benjamin-Cummings, Menlo Park
Wong JTF (1975) A co-evolution theory of the genetic code. Proc Natl Acad Sci U S A 72:1909–1912
Wong JTF, Ng SK, Mat WK et al (2016) Coevolution theory of the genetic code at age forty: pathway to translation and synthetic life. Life (Basel) 6:12
Yarus M (1998) Amino acids as RNA ligands: a direct-RNA-template theory for the code's origin. J Mol Evol 47:109–117
Yarus M (2017) The genetic code and RNA-amino acid affinities. Life 7:13

Chapter 8
The Origin of Gene

Abstract There are two sides of the origin of gene. The first one (the origin of gene-I) is to understand the way how the first gene was created through random process on the primitive Earth. The second one (the origin of gene-II) is to clarify the way how an entirely new (EntNew) gene or the first gene of a gene family has been created from the past to the present. The origin and evolutionary process of gene could not be rationally explained by the gene-duplication theory (Ohno S. Evolution by gene duplication. Springer, Heiderberg, 1970) and the exon-shuffling theory (Gilbert W. Cold Spring Harb Symp Quant Biol 52:901–905, 1987). On the contrary, it would be reasonably explained by anticodon joining hypothesis, how the first gene was created through random process as follows. Genetic sequence for protein synthesis is written with not mere base sequence but codon sequence. On the other hand, the first gene must be created through random process on the primitive Earth. This means that the first gene must be created by random joining of codons or anticodons, and that the only one way for creation of the first gene would be to randomly join anticodons carried by the first four nonspecific AntiC-SL tRNAs according to the "anticodon joining hypothesis" on the origin of gene-I, because there existed no codon anywhere on the primitive Earth.

I consider about the origin of gene-II on creation of EntNew genes as follows. Understanding the mechanism, how an EntNew gene/protein or the first ancestral gene/protein of a family was created, should be one of the most important issues in the biological sciences. However, the mechanism is also totally unknown still now. On the other hand, creation of such a mature EntNew gene/protein should be, of course, carried out through a random process, because it cannot be designed in advance. However, both an EntNew gene and an EntNew protein never be created by random polymerization of nucleotides and amino acids, because of the respective extraordinary large sequence diversities of $\sim 10^{180}$ and $\sim 10^{130}$. Protein 0th-order structure or a specific amino acid composition, in which immature but water-soluble globular protein can be produced even through random process, holds the key for solving the difficult problem of creation of an EntNew gene/protein. I describe three processes generating EntNew genes under the three genetic codes or the universal genetic code, SNS primitive genetic code and GNC primeval genetic code in this chapter, and discuss why the EntNew gene encoding a mature protein could be created through the respective random processes.

K. Ikehara, *Towards Revealing the Origin of Life*,
https://doi.org/10.1007/978-3-030-71087-3_8

Keywords Origin of gene · Gene-duplication theory · Exon-shuffling theory · Self-replicated RNA theory · Anticodon joining hypothesis · Pan-GC-NSF(a) hypothesis

How was the first gene created through random process on the primitive Earth? How entirely new gene has been generated from the past to the present? The answer for solving the riddle of the origin of gene is described in this chapter

1 Introduction: Towards Solving the Riddle of the Origin of Gene

First, it is described in this section, what matters are necessary to explore and to solve the riddle of the origin of gene, as similarly to the other chapters so far Based on the matters, it is discussed how modern genes have been evolved from the most primitive genes, which were created on the primitive Earth (Figs. 8.1 and 8.2).

1.1 Necessary Matters for Exploring the Origin of Gene

1. It is necessary to well understand what features modern gene has, in order to make clear the origin of gene composing the fundamental life system, so that it can be confirmed, whether or not the idea for solving the origin of gene assumed is valid, by comparison of the features of the first gene assumed with those of modern gene as a target.
2. By doing so, it can be confirmed from what kind of primitive gene contemporary genes have originated.
3. If it must be considered that the contemporary gene originated from a primitive one with features different from the contemporary one, as a matter of course, the conversion process from the primitive gene assumed to the first gene with similar features to modern one must be reasonably explained, either directly or indirectly.
4. Furthermore, it must be confirmed whether or not an evolutionary process from the primitive gene, which could be created by a random process on the primitive Earth, to modern one.

Fig. 8.1 Base sequence (small letters: upper lines) of *Escherichia coli rpoZ* gene (Genbank; JW3624) and amino acid sequence (capital letters: lower lines) of RNA polymerase, omega subunit encoded by the *rpoZ* gene

```
atggcacgcgtaactgttcaggacgctgta
M  A  R  V  T  V  Q  D  A  V
gagaaaattggtaaccgttttgacctggta
E  K  I  G  N  R  F  D  L  V
ctggtcgccgcgcgtcgcgctcgtcagatg
L  V  A  A  R  R  A  R  Q  M
caggtaggcggaaaggatccgctggtaccg
Q  V  G  G  K  D  P  L  V  P
gaagaaaacgataaaaccactgtaatcgcg
E  E  N  D  K  T  T  V  I  A
ctgcgcgaaatcgaagaaggtctgatcaac
L  R  E  I  E  E  G  L  I  N
aaccagatcctcgacgttcgcgaacgccag
N  Q  I  L  D  V  R  E  R  Q
gaacagcaagagcaggaagccgctgaatta
E  Q  Q  E  Q  E  A  A  E  L
caagccgttaccgctattgctgaaggtcgt
Q  A  V  T  A  I  A  E  G  R
cgttaa
R  *
```

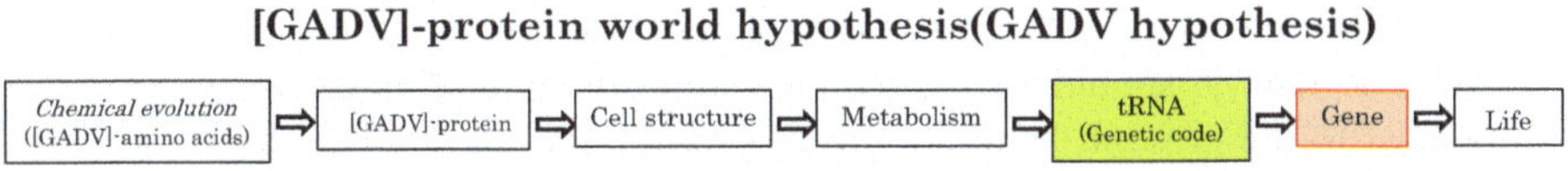

Fig. 8.2 In this chapter, the origin of gene is discussed (red box). The first single-stranded RNA, which led to generation of the first double-stranded RNA gene, was created by random joining of anticodons, GNCs, carried by four non-specific anticodon stem-loop (AntiC-SL) tRNAs (see also Chap. 6). Therefore, it can be understood that formation of the first gene was assisted by the previous formation of the primitive tRNAs (yellow box). It simultaneously indicates that gene never be produced directly by random joining of nucleotides. Acquisition of the first double-stranded RNA gene, which was produced by complementary strand synthesis of the single-stranded $(GNC)_n$ RNA and matured, triggered prosperity of diverse extant organisms on this planet

1.2 General Structural Features of Gene (Basic Features of Gene)

1. Gene is composed of not base sequence but codon sequence for encoding a mature protein (Table 8.1).
2. Genetic information is written as a codon sequence on double-stranded RNA or DNA, so that one strand carries genetic information and the other strand can function as a template for gene expression. The codon sequence (nonstop frame) on antisense strand of a GC-rich gene (GC-NSF(a)) can supply the field for EntNew gene creation.

1.3 General Functional Features of Gene

1. Gene generally carries information for synthesis of an amino acid sequence of a mature protein.

1.4 There Are Two Sides for Exploring the Origin of Gene

Next, I would like to explain that there are two sides in the origin of gene or there are two origins of gene-I and gene-II, before the origin of gene, which I have considered, is explained.

1. The origin of gene-I: How was the first gene encoding a mature protein created on the primitive Earth? The studies on the origin of gene-I usually aim at this problem.
2. The second origin of gene-II: EntNew genes, which do not show any meaningful homology with every previously existing gene and, therefore, does not belong to any gene family, should have been continuously created when necessary from the past to the present. In the origin of gene-II, the mechanism, how and from where an EntNew gene has been created, is studied.

Table 8.1 Necessary conditions that should be kept in mind when considering creation of the first gene

1. Genetic information is composed of not base sequence but codon sequence.
2. Codon, of course, did exist nowhere on the primitive Earth.
3. Therefore, anticodons carried by AntiC-SL tRNAs had to be arranged to create the first codon sequence.
4. The first codons must be symmetric, like GNC, so that EntNew gene can be created from GC-$(GNC)_n$-NSF(a).

Gene is intimately related to protein through genetic expression. Therefore, many figures used to explain the origins of gene-I and -II duplicate especially between Chap. 3 and this chapter. Hence, some figures are omitted in this chapter and the places of the figures appeared in other chapters are indicated in the text.

2 The Ideas on the Origin of Gene Advocated by Other Researchers

2.A. The Origin of Gene-I: How Was the First Gene Created?
To the best of my knowledge, any concrete idea about this problem has not been presented by other researchers.

2.B. The Origin of Gene-II: How Were Entirely New Genes Created from the Past to the Present?
Two main theories have been proposed as ideas for explaining the origin of gene-II thus far. Then, it is discussed whether or not the two theories are proper for the origin of gene-II based on the necessary matters and general features of modern gene described in Sects. 1.1, 1.2 and 1.3, although the two theories have not always been presented for solving the origin of gene-II and the creation mechanism of EntNew genes.

2.1 Gene-Duplication Theory

Gens are generally classified as gene families and superfamilies, according to similarities of among base sequences or amino acid sequences and to catalytic properties of proteins encoded by the genes. Therefore, a gene family is composed of genes with similar base sequences encoding proteins with a similar function, which originated from one ancestor gene, whereas a gene superfamily contains several gene families encoding proteins with similar amino acid sequences but with different catalytic functions. Therefore, it is considered that not only genes belonging to a gene family in a narrow sense but also those in a superfamily originated from one ancestor gene. This indicates that all genes except the most ancestral genes in the gene families and the gene superfamilies were produced according to the "gene-duplication theory", which was proposed by S. Ohno about 50 years ago (Ohno 1970).

The theory suggests that a new gene is produced from one of duplicated genes. Thereafter, it has been confirmed that the theory can well explain the way how new homologous genes in a gene family and a gene superfamily have been created. However, the gene-duplication theory has a fatal weakness that a previously existed gene is always required to create a new homologous gene. Therefore, the theory cannot show the way how an EntNew gene or the first family gene was created.

It would be obvious that a codon sequence encoding a mature protein cannot be designed in advance, and, therefore, an EntNew gene must be produced by random polymerization of nucleotides. However, an EntNew gene could not be produced by random joining of nucleotides, because the diversity of codon sequence encoding a small protein composed of 100 amino acids is also extraordinary large as $(4^3)^{100} = {\sim}10^{180}$ (Chap. 3: Fig. 3.4). Furthermore, the genetic information for a mature protein synthesis could not be generated without a target of protein (Chap. 3: Sect. 4.8.2). Therefore, it is totally unknown how an EntNew gene is created, regardless of one of the significant matters in biological sciences. Those would be the reason, why formation process of protein always starts from amino acid sequence in every textbook still now (Chap. 3: Fig. 3.3). Then, discuss in this chapter, the way how the origin of gene-II encoding a mature protein could be solved (see also Chap. 3: Sect. 4.8).

Strengths of Gene-Duplication Theory The gene-duplication theory well explains the process for forming various new homologous genes in a gene family and a gene superfamily. Therefore, genes formed under the theory are compatible with the general features of gene.

Weaknesses of Gene-Duplication Theory However, the gene-duplication theory has a fatal weakness that a previously existing gene is always required to create a new homologous gene. Therefore, the theory cannot show the creation process, how an EntNew gene or the first family gene was created, which is one of the most important points for solving the origin of life.

2.2 Exon-Shuffling Theory

The "exon shuffling theory" was proposed by Gilbert (1987), assuming that a new functional gene is created from a plural number of exons shuffled among exons of several genes. However, the exon shuffling theory also requires a plural number of previously existed genes. Therefore, the theory cannot explain the way how an original ancestor gene or an EntNew gene, which is quite different from any previously existed genes, was and have been created.

Strengths of Exon-Shuffling Theory The exon-shuffling theory would be one of influential ideas, by which formation process of new genes could be explained. It is considered that the exon-shuffling theory is a gene-version of the short peptide (or motif) hypothesis on the origin of protein (Romero et al. 2016) (Chap. 3: Sect. 3.3), because a new protein could be formed by combination of short peptides indirectly through exon-shuffling (the exon-shuffling theory) or directly (the short peptide theory).

Weaknesses of Exon-Shuffling Theory However, it would be difficult to construct new proteins according to the exon-suffling theory with a small number of exceptions, because a mature protein could not be constructed by combination of partial structures (peptides) produced by expression of exons, as explained in Chap. 3: Sect. 3.3, and also because short peptides or motif could not retain the original structures in a new protein or in a new environment, which was formed by assembly of short peptides from several different proteins. In addition, the theory does not also explain the most fundamental problem on the creation of the first gene or an EntNew gene. Therefore, the exon-shuffling theory also could not contribute to elucidation of the origin of gene-II, similarly to the gene-duplication theory.

2.3 Self-Replicated RNA Theory (RNA World Hypothesis)

In the RNA world hypothesis, which was proposed for explaining the process to the emergence of life, it is assumed that self-replicated RNA could acquire a genetic information for protein synthesis one day and a protein could be produced owing to the genetic information (Gilbert 1986). Therefore, I named the idea, that RNA gene originated from the self-replicated RNA, as "self-replicated RNA hypothesis" for convenience of discussion in this chapter.

Strengths of Self-Replicated RNA Theory The RNA world hypothesis was originally presented to explain the origin of life. Therefore, any concrete idea, about the mechanism, how EntNew genes could be created with the self-replicated RNAs, has not been presented thus far, because the theory or the idea was proposed to explain only the process from appearance of self-replicated RNA to the emergence of life. Therefore, there is no specific strength on the origin of gene-II in the theory.

Weaknesses of Self-Replicated RNA Theory In addition, RNA obtained by self-replication or by random joining of nucleotides (although, I cannot even imagine that RNA could be produced with prebiotic means and can be self-replicated) should be composed of a random base sequence but not codon sequence, therefore, the self-replicated RNA should be only a chemical material without any genetic information for protein synthesis (Chap. 3: Fig. 3.5), meaning that the RNA does not satisfy the necessary condition 1 of Table 8.1. In addition, tRNA, which is indispensable for the self-replicated RNA or RNA gene to express the genetic information, should not exist at the time point, when self-replicated RNA was produced. This point also indicates that the self-replicated RNA should be meaningless for the protein synthesis.

Thus, I must conclude here that the origin of gene-II cannot be explained rationally by the three ideas proposed by other researchers. Next, I present my idea explaining the ways how the first gene and EntNew genes were created.

3 My Idea About the Origin of Gene

3.1 For What Purpose Was the First Genetic Information (Gene) Created?

Genetic information (gene) codes for an amino acid sequence, which is necessary to synthesize a mature protein. Could such a gene be first created among the three members of the core life system (Chap. 2: Fig. 2.1a), or gene, tRNA (genetic code) and protein? It would be impossible to create genetic information for protein synthesis independently of protein, because information itself could not be generated without a concrete object. These indicate that the first gene must be created after protein and/or tRNA was formed. Then, it is discussed in the following sections how the first gene (genetic information) was created.

I present two my own ideas in this chapter for explaining the origin of gene-I or "How was the first gene created?"; anticodon joining hypothesis, and for explaining the origin of gene-II or "How were entirely new genes created from the past to the present?"; pan-GC-NSF(a) hypothesis.

3.A. The Origin of Gene-I: How Was the First Gene Created?

Here, I discuss the origin of gene-I or explain how the first gene was created. For the purpose, it must be kept in mind that the first gene, of course, must be produced through a random process, only which proceeded on the primitive Earth. After a double-stranded (ds)-$(GNC)_n$ RNA could be formed for the first time on the primitive Earth, the ds-RNA became the first $(GNC)_n$ gene through evolutionary process from a $(GNC)_n$ codon sequence coding for an immature [GADV]-protein to a mature gene encoding a mature [GADV]-protein with the two general features of modern gene (Sects. 1.2 and 1.3). I will also consider with those in mind in this section, whether or not my "anticodon joining hypothesis" is reasonable as an idea for explaining the origin of gene-I.

First, I will explain three preliminary steps for the creation of the first gene in short in order to avoid overlapping description with Chap. 3: Sect. 4.5, before entering into discussion of my idea, the anticodon joining hypothesis on the origin of gene-I (Fig. 8.3).

Step 1 (Fig. 8.4): [GADV]-proteins could be synthesized by direct random joining of [GADV]-amino acids, owing to the [GADV]-protein 0th-order structure (Chap. 3: Sect. 4.5.1).

Step 2 (Fig. 8.4): At the next step, [GADV]-proteins with random amino acid sequence could be produced with [GADV]-AMPs or activated [GADV]-amino acids and with [GADV]-aa-ACC more efficiently than before (Chap. 3: Sect. 4.5.2).

Step 3 (Fig. 8.4): Successively, random [GADV]-proteins could be produced with four nonspecific [GADV]-AntiC-SL-tRNAs carrying 5′-CCA-3′ end (Chap. 3: Sect. 4.5.4).

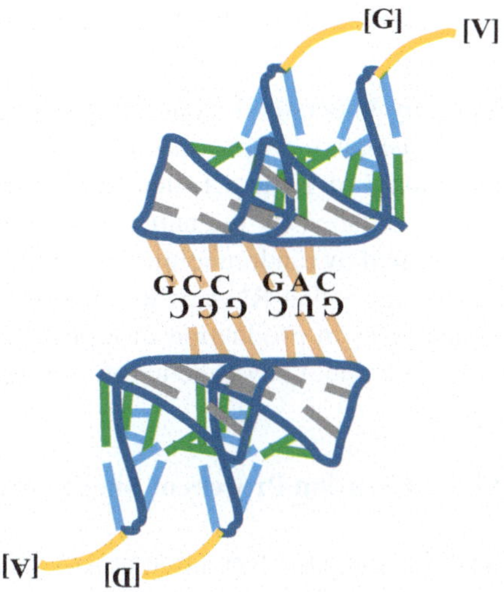

Fig. 8.3 Formation of two dimers (as a whole, a tetramer), which were composed of two AntiC-SL tRNAs bound vertically through base pairs between two anticodons. The tetramer formation of AntiC-SL tRNAs made it possible to create the first genetic information, because two anticodons could be positioned side by side to form phosphodiester bond between the two anticodons for the first time. The AntiC-SL was drawn by tracing the AntiC-SL of modern Yeast Phe-tRNA (PDBj, (1EHZ))

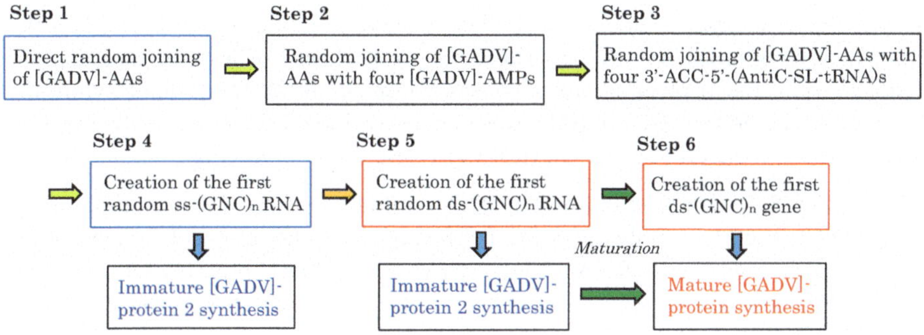

Fig. 8.4 Creation process of the first double-stranded $(GNC)_n$ gene. [GADV]-proteins (actually aggregates of [GADV]-peptides) were synthesized Step 1: by direct random joining of [GADV]-amino acids ([GADV]-AAs). Step 2: with activated [GADV]-AA or four [GADV]-aminoacyl AMPs and, Step 3: through [GADV]-aa-ACCs followed by [GADV]-aa-ACC-AntiC-SL tRNAs. Successively, Step 4: the first single-stranded (ss)-RNA with random $(GNC)_n$ sequence was created by random joining of anticodons, GNCs, in the AntiC-SLs. Step 5: the first double-stranded (ds)-RNA with random $(GNC)_n$ sequence on both strands was produced by complementary strand synthesis of the ss-$(GNC)_n$ RNA (bold orange arrow). Finally, Step 6: the first ds-$(GNC)_n$ RNA gene was created through maturation of ds-RNA with random $(GNC)_n$ sequences. Bold blue arrows and bold green arrows indicate gene expression and maturation from an immature to a mature EntNew gene/protein, respectively

3.2 Anticodon Joining Hypothesis

As easily understood from the first general features of gene (Sects. 1.2 and 1.3), gene must carry not base sequence but codon sequence. Furthermore, the necessary condition 3 of Table 8.1, that tRNA must be formed before the first gene was created, is satisfied by the anticodon joining hypothesis assuming that the first gene was created by random joining of anticodons carried by AntiC-SL tRNAs (Fig. 8.3). It also means that RNA, which was produced before tRNA was formed, would be meaningless in production of a protein and, therefore, would be a mere chemical material without any genetic information.

3.2.1 Creation Process of Single-Stranded $(GNC)_n$ RNA

Next, I explain the formation process of single-stranded (ss)-$(GNC)_n$ RNA based on the AntiC-SL tRNA hypothesis on the origin of tRNA. As described in the first condition of gene (Table 8.1), not bases (actually, nucleotides) but codons must be arranged in even the first RNA to produce an immature [GADV]-protein. However, as a matter of course, codons never be laid around on the primitive Earth. Not the case, it is recognized for the first time that triplet bases of a base sequence are codons, when the triplets are read by anticodon of tRNA for protein synthesis. Creation of the first ss-$(GNC)_n$ RNA was made possible for the first time by acquisition of four non-specific AntiC-SL tRNAs as described below (Figs. 8.3 and 8.4).

Step 4. (Fig. 8.4): The first ss-$(GNC)_n$ RNA could be generated by random joining of GNC anticodons embedded into the respective AntiC-loops of four primeval nonspecific AntiC-SL tRNAs. Thus, immature [GADV]-proteins with random [GADV]-amino acid sequence could be produced by translation of the ss-$(GNC)_n$ RNA (Chap. 6: Fig. 6.9).

Therefore, it means that codon sequence must be formed by joining of anticodons carried by the four AntiC-SL tRNAs. In other words, only one way for creating the first genetic information is to join anticodons of the two AntiC-SL tRNAs randomly juxtaposed (Fig. 8.3) (see also Chap. 6: Fig. 6.12). Then, it is favorable that anticodon is symmetric like as GNC for creation of a codon sequence through base pairs formed between two anticodons (Fig. 8.3). The primeval genetic code, GNC, fortunately satisfies the condition too.

Two dimers, which were formed by juxtaposition of two AntiC-SL tRNAs bound vertically, must be closely positioned to create the first $(GNC)_n$ codon sequence through joining two anticodons carried by the two AntiC-SL tRNAs with phosphodiester bond (Fig. 8.3) (see also Chap. 6: Fig. 6.12). From the fact that two codons closely arranged on mRNA can be read by two anticodons of modern tRNAs, which are much larger than the most primitive AntiC-SL tRNA, it can be easily understood that two anticodons of two AntiC-SL tRNAs adjoined could be positioned to form easily phosphodiester bond.

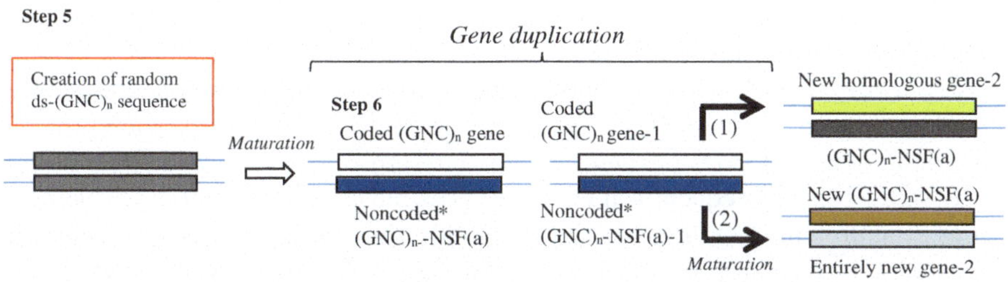

Fig. 8.5 After duplication of the first ds-(GNC)$_n$ gene encoding a mature [GADV]-protein, a new homologous gene (1) and another EntNew gene (2) could be generated, whenever a necessity arose, from the sense strand and the antisense strand of the first ds-(GNC)$_n$ gene, respectively. The "noncoded" is used as the word meaning that not a mature but an immature protein is synthesized from the sequence on antisense strand. The mechanism generating an EntNew gene/protein is essentially the same under the three genetic code or GNC code, SNS code and the universal genetic code. Therefore, the formation process of EntNew gene can be understood by replacing GC-rich (GNC)$_n$ gene and GC-(GNC)$_n$-NSF(a) with GC-rich (SNS)$_n$ gene and GC-(SNS)$_n$-NSF(a), and with GC-rich gene and GC-NSF(a), respectively

However, the first ss-(GNC)$_n$ RNA formed by joining of GNC codons was not a RNA with a large number of codons. Not the case, the ss-(GNC)$_n$ RNA would be still much smaller than mRNA encoding a polypeptide with a large number of [GADV]-amino acids. Therefore, only [GADV]-peptides could be produced by translation of the first ss-(GNC)$_n$ RNA.

3.2.2 Formation of Double-Stranded (GNC)$_n$ RNA

Step 5 (Figs. 8.4 and 8.5): Successively, the first ds-(GNC)$_n$ RNA encoding an immature [GADV]-protein was formed by complementary strand synthesis of the ss-(GNC)$_n$ RNA. However, any mature protein could not be still synthesized with the ds-(GNC)$_n$ RNA, because one sequence of the ds-RNA was created by random joining of anticodons carried by tRNAs and the other sequence was formed by complementary strand synthesis of the ss-RNA with random (GNC)$_n$ sequence. Therefore, [GADV]-protein produced from either one of two strands of the ds-(GNC)$_n$ RNA was only incomplete and immature at the time point.

3.2.3 Creation of the First Double-Stranded (GNC)$_n$ RNA Gene Encoding a Mature [GADV]-Protein

Therefore, any mature protein could not be produced by expression of the first ds-(GNC)$_n$ RNA, because the situation of the [GADV]-protein synthesis is essentially the same as random and direct joining of [GADV]-amino acids. That is, even if an immature protein could be produced by expression of the ds-RNA, the ds-(GNC)$_n$

RNA could not code for any genetic information, because a gene must encode a mature protein (Sect. 1.3).

Step 6 (Fig. 8.5): Thus, one strand became the first sense strand and the other strand became an antisense strand or a template strand through maturation process from an immature [GADV]-protein. Asymmetry arose between two strands of the ds-$(GNC)_n$ RNA. Consequently, it was made possible to generate both a new homologous gene/protein from the sense sequence and another EntNew gene/protein from the antisense sequence of the first ds-$(GNC)_n$ gene, respectively, after gene-duplication when necessary (Fig. 8.5).

Only one way for overcoming the difficulty of creation of the first gene is to first synthesize an immature protein with random [GADV]-amino acids through expression of the ds-$(GNC)_n$ RNA with random codon sequences and, successively, to evolve the immature protein with memorizing ability of the ds-$(GNC)_n$ RNA for amino acid replacements to a mature protein (refer also Chap. 3: Fig. 3.12). Thus, the first ds-$(GNC)_n$ gene encoding a mature [GADV]-protein like a precision polymer machine could be produced through maturation of the immature gene for the first time (Figs. 8.4 and 8.5: Step 6). Furthermore, note that the appearance order of immature [GADV]-protein, AntiC-SL tRNA and $(GNC)_n$ gene is also supported by the origin of metabolism, which started from syntheses of [GADV]-amino acid followed by nucleotide synthesis (see Chap. 5: Figs. 5.7 and 5.8).

Accompanied by the appearance of ds-$(GNC)_n$ gene, it seems like that the leading role of protein in the core life system composed of gene, tRNA (genetic code) and protein (Chap. 3: Fig. 3.3), was transferred to gene. However, it should be recognized that extant organisms are living chiefly using proteins, which can exhibit various functions necessary to live, and gene play only auxiliary role in carrying genetic information. Therefore, protein plays essentially the leading role even in the present fundamental life system. This means that life, which emerged from the "protein world"on the primitive Earth, are still living not in the "DNA/RNA or gene world" but in the "protein world". This also clearly indicates that RNA-centered world did not exist on this planet anytime from the past to the present.

Strengths of Anticodon Joining Hypothesis It could be concluded according to the necessary conditions for the origin of something that the anticodon joining hypothesis is valid as the origin of gene-I, because it can be explained that the most primitive $(GNC)_n$ gene was created through random process, and because the evolutionary process from the first $(GNC)_n$ gene to modern genes can be also reasonably explained, as is described in detail in the next Sect. 3.B.

I consider further that it would be probably impossible to explain the creation processes of the three members, gene, tRNA (genetic code) and protein, with other ideas than GADV hypothesis.

Weaknesses of Anticodon Joining Hypothesis The anticodon joining hypothesis on the origin of gene-I was proposed based on the AntiC-SL tRNA hypothesis on the origin of tRNA, which was obtained by sequence analysis of 5′ stem sequences of AntiC-SLs of *Pseudomonas aeruginosa* tRNAs. However, it is also a great weakness in the anticodon joining hypothesis that it is not confirmed by experiments whether or not the first $(GNC)_n$ RNA gene could be really created through the process as shown in Figs. 8.4 and 8.5. Although, of course, that thing does not always mean that the ss-$(GNC)_n$ RNA cannot be produced by random joining of anticodons of AntiC-SL tRNAs, the first ds-$(GNC)_n$ RNA cannot be created by complementary strand synthesis of the first ss-$(GNC)_n$ RNA, and the first ds-$(GNC)_n$ gene cannot be formed through the maturation process of ds-$(GNC)_n$ RNA encoding an immature [GADV]-protein.

3.B. The Origin of Gene-II: How Were Entirely New Genes Created from the Past to the Present?

As a matter of course, EntNew genes have been always created even after the creation of the first $(GNC)_n$ gene when necessary. In this section, first, the mechanism, how the EntNew genes were created under GNC genetic code, is explained and discussed.

3.3 Creation of an Entirely New Gene After Formation of the First Double-Stranded $(GNC)_n$ RNA Gene

3.3.1 Creation of an Entirely New Gene Under GNC Primeval Genetic Code (GC-$(GNC)_n$-NSF(a) Hypothesis)

Every modern enzyme generally recognizes one organic compound or a substrate and excludes any other organic compounds (Chap. 3: Sect. 1.3). Such a mature protein always must be created from an immature water-soluble globular protein with some flexibility, which were produced by expression of a GC-NSF(a) (Fig. 8.6, Table 8.2) (Ikehara et al. 1996; Ikehara 2002), because it is impossible to directly create any mature protein through random process (see also Chap. 3: Sect. 4.3) (Ikehara 2014).

The two conditions described below must be satisfied in order to generate an EntNew $(GNC)_n$ gene encoding a mature protein under the GNC primeval genetic code (Table 8.2). (1) Use of ds-$(GNC)_n$ RNA memorizing amino acid replacements during evolution from an immature [GADV]-protein-2 to a mature [GADV]-protein (Chap. 3: Figs. 3.11 and 3.12). (2) An EntNew gene encoding a mature [GADV]-protein with an ordered sequence must be created from a GC-$(GNC)_n$-NSF(a) encoding an immature water-soluble globular [GADV]-protein-2 with an essentially random amino acid sequence (Chap. 3, Figs. 3.11 and 3.12). Therefore, the creation process of such a mature protein from an immature protein under GNC code is as follows.

(1) An immature protein-2, which was produced by expression of a GC-$(GNC)_n$-NSF(a), could evolve to a mature protein, if the immature protein-2 exhibited even a weak catalytic activity, which was necessary for the life to live. Therefore,

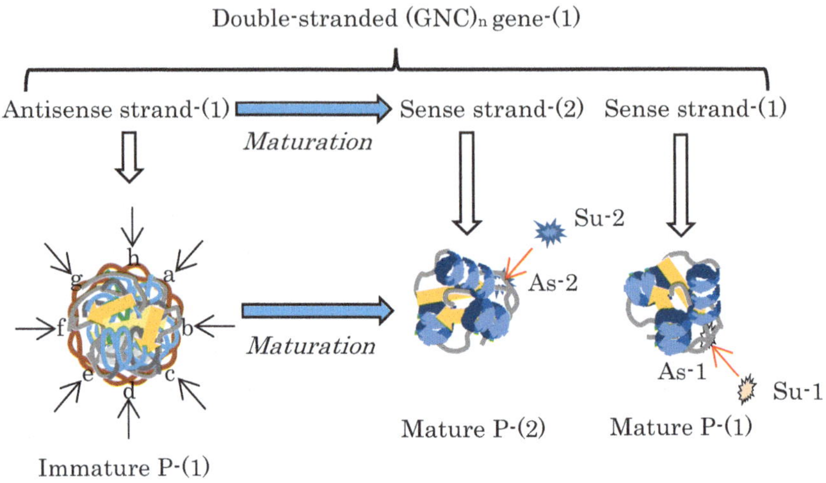

Fig. 8.6 Creation of a new mature P-(2) from an immature P-(1), which was produced from antisense strand-(1) of GC-rich $(GNC)_n$ RNA gene-(1). The memorizing ability of base substitutions of the ds-RNA made it possible to evolve the immature P-(1) to a mature P-(2) with rigid and compact structure, as raising function of the active site (As) for the substrate (Su) and, in parallel, other unnecessary catalytic activities are excluded. Numerals (1) and (2) indicate the first and the second generations for formation of the mature gene/protein, respectively

Table 8.2 Two conditions for producing of entirely new [GADV]-proteins after creation of the first $(GNC)_n$ gene. The reason, why the immature protein must be utilized for the creation of a mature protein, is because amino acid sequence of the mature protein cannot be designed previously

1. Memorizing ability of a double-stranded RNA gene must be utilized to create of a mature protein.
2. The mature [GADV]-protein must be created though evolution of an immature [GADV]-protein produced from a GC-$(GNC)_n$-NSF(a), which encodes substantially random [GADV]-amino acid sequence taken out from the protein 0th-order structure.

an EntNew $(GNC)_n$ gene can be created from a GC-$(GNC)_n$-NSF(a) as raising the activity of immature [GADV]-protein-2, because the immature [GADV]-protein-2 could evolve to a mature protein by using the ability of the ds-$(GNC)_n$ gene memorizing both base substitutions and amino acid replacements (Fig. 8.6). Upon the evolution of the immature [GADV]-protein to a mature protein, the structure of the immature protein became rigid to be able to refine the protein structure and reject any other organic compound. Simultaneously, the original mature $(GNC)_n$ gene-(1) and the immature $(GNC)_n$-NSF(a)-(1) became immature $(GNC)_n$-NSF(a)-(2) and a new mature $(GNC)_n$ gene-(2), respectively (Fig. 8.6). Numeral in parenthesis means the respective generation of mature protein formation.

(2) Thus, a new homologous protein and an EntNew protein could be produced from two $(GNC)_n$ codon sequences on sense strand and from the other $(GNC)_n$ sequence on antisense strand of ds-$(GNC)_n$ gene, respectively (Fig. 8.5). By

> repeating the same processes, various new homologous genes and EntNew genes could be created to be able to produce new homologous proteins and EntNew [GADV]-proteins, respectively (Figs. 8.5 and 8.6).

Thus, many EntNew mature genes have been created as a trigger of a polypeptide produced by expression of GC-$(GNC)_n$-NSF(a) without interruption until now, after creation of the first ds-$(GNC)_n$ RNA gene, because even the polypeptide chain with an essentially random amino acid sequence could be folded into a water-soluble structure or an immature protein with some flexibility (Fig. 8.6). Of course, strictly stating, the [GADV]-amino acid sequence obtained by expression of a GC-$(GNC)_n$-NSF(a) was not completely random, because the sequence is specified and restricted by the sense sequence of the corresponding $(GNC)_n$ gene. Nevertheless, the amino acid sequence produced by expression of $(GNC)_n$-NSF(a) can be regarded as substantially random, owing to the anti-parallel ds-$(GNC)_n$ RNA structure (Fig. 8.7) (Ikehara 2016a).

Note that both (1) production of immature [GADV]-protein-1 by random direct joining of [GADV]-amino acids in the absence of gene, and (2) synthesis of immature [GADV]-protein-2 from antisense codon sequence of ds-$(GNC)_n$ RNA gene rely on the protein 0th-order structure. Therefore, both of the pluripotent immature protein-1 and -2 could be formed through the two different routes, respectively (Chap. 3: Fig. 3.12).

It is also important to recognize that an organic compound led the creation of an EntNew protein catalyzing the organic compound or substrate (see also Chap. 3: Sect. 4.8). The reasons are because an active site fitting to the substrate never be

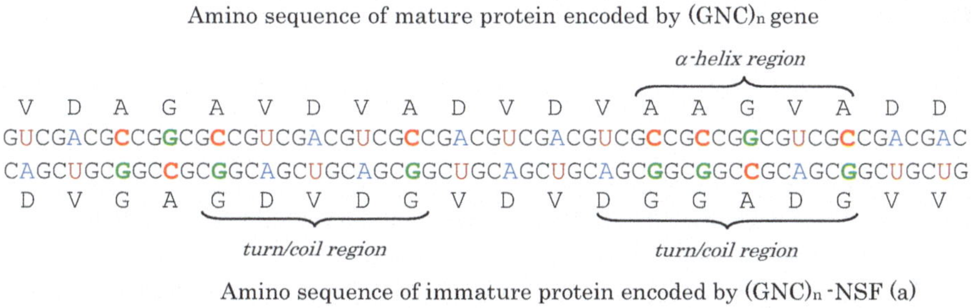

Fig. 8.7 Partial amino acid sequences of a mature protein encoded by $(GNC)_n$ gene (upper base sequence) and of an immature protein encoded by $(GNC)_n$-NSF(a) (lower base sequence). Letters written at the second codon position with colored alphabets indicate four bases, green G: guanine, red C: cytosine, blue A: adenine, and brown U: uracil. Six Cs (30%), As (30%), Us (30%) and two Gs (10%) are randomly arranged at the second codon position of a partial $(GNC)_n$ gene, taking the results of computer analysis (C_2 (31%), A_2 (29%), U_2 (29%) and G_2 (11%)) shown in Chap. 7; Fig. 7.6 into consideration (Ikehara et al. 2002). It can be seen that two amino acid sequences encoded by $(GNC)_n$ gene and $(GNC)_n$-NSF(a) are quite different from each other. The difference is caused by anti-parallel ds-RNA structure and by translation under the GNC code. Consequently, polypeptide chain encoded by codon sequence on antisense strand is folded into turn/coil conformation at a higher probability causing more flexible than that encoded by sense codon sequence

created at one stroke and must be always formed through evolutionary process from an incomplete one to the complete one as accumulating amino acid replacements and as led by the substrate, step by step (Fig. 8.6, Chap. 3: Fig. 3.12). That is, the mature enzyme catalyzing a chemical reaction never be created without the organic compound. Similarly, every EntNew gene encoding a mature protein never be created without the corresponding immature protein, because the creation of the EntNew gene is always lead by functional improvement of the protein (see also Chap. 3: Sect. 4.8).

3.3.2 An Immature [GADV]-Protein-2 is More Flexible Than a Mature Protein

An immature pluripotent [GADV]-protein-2 (iPP-2), which is produced by expression of antisense sequence of a $(GNC)_n$ gene, is more flexible than a mature [GADV]-protein (Chap. 3; Figs. 3.12 and 3.13). The flexibility of the immature protein contributed to generation of a large number of possible catalytic sites (pluripotency) leading to creation of a mature protein (Chap. 3, Sect. 4.8).

A mature [GADV]-protein produced from a $(GNC)_n$ gene satisfies the four conditions for protein structure formation, when Ala encoded by GCC codon was used more frequently than Gly encoded by GGC codon in the protein (Chap. 3, Fig. 3.10; Chap. 7, Fig. 7.6) (Ikehara et al. 2002; Ikehara 2002). In other words, the [GADV]-protein could take the structure similar to extant proteins, when α-helix forming amino acid, Ala, was used considerably at a higher frequency than turn/coil forming amino acid, Gly. This indicates that an immature [GADV]-protein synthesized from a $(GNC)_n$-NSF(a) must be more flexible than the mature [GADV]-protein, because GCC codons for Ala on sense strand are replaced by GGC codons for Gly on antisense strand (Fig. 8.7). Therefore, the contents of Ala and Gly in a mature [GADV]-protein encoded by a gene on sense strand, are equal to Gly and Ala contents in the corresponding immature [GADV]-protein encoded by the $(GNC)_n$-NSF(a), respectively (Fig. 8.7). It means that the structure of the immature [GADV]-protein was more flexible than the mature protein. The flexible structure enabled to generate a large number of premature catalytic sites on the immature [GADV]-protein to easily adjust to a newly encountered an organic compound.

Thus, the excellent assignment of four [GADV]-amino acids into four GNC codons in the GNC code made it possible to create efficiently an EntNew $(GNC)_n$ gene encoding an EntNew mature [GADV]-protein.

3.3.3 Creation of an Entirely New Gene Under SNS Primitive Genetic Code (GC-$(SNS)_n$-NSF(a) Hypothesis)

Mechanism generating an EntNew gene/protein under SNS genetic code encoding 10 amino acids is essentially the same with that under the GNC code (Fig. 8.5) (Ikehara and Yoshida 1998; Ikehara 2002). Therefore, the mechanism used in the

SNS code era is described as focusing on only main points in this section, because their details are described in the previous Sects. 3.3.1 and 3.3.2. Hence, the formation process of EntNew gene can be understood by replacing $(GNC)_n$ gene and $(GNC)_n$-NSF(a) with $(SNS)_n$ gene and $(SNS)_n$-NSF(a), respectively, except the first step of "creation of random ds-$(GNC)_n$ sequence" (Fig. 8.5).

(1) First, one GC-$(SNS)_n$ gene-(1) was duplicated into two. Original genetic function was conserved in one of the duplicated genes. Therefore, multiple mutations could be accepted on the other duplicated GC-$(SNS)_n$-NSF(a)-(1) to create an EntNew gene.
(2) An immature protein-(1) could be produced by expression of the GC-$(SNS)_n$-NSF(a)-(1) (Fig. 8.5).
(3) If one catalytic function was found on the surface of the immature protein-1, the catalytic activity could be elevated as receiving multiple base substitutions (Fig. 8.6). In parallel, structure of the immature protein gradually became rigid and compact accompanied by maturation to a mature protein. Simultaneously, the original mature gene-(1) and the immature GC-$(SNS)_n$-NSF(a)-(1) became an immature GC-$(SNS)_n$-NSF(a)-(2) and a mature GC-$(SNS)_n$ gene-(2), respectively (Fig. 8.5).

Thus, the reason, why an EntNew gene/protein could be generated from nonstop frame on antisense strand of a $(SNS)_n$ gene (GC-$(SNS)_n$-NSF(a)) under the SNS genetic code too, is because GC-$(SNS)_n$-NSF(a) could code for an essentially random amino acid sequence of an immature water-soluble globular protein with some flexibility, as similarly as the case of GC-$(GNC)_n$-NSF(a) and a modern GC-NSF(a) (Fig. 8.5). The reasons, why GC-$(SNS)_n$-NSF(a) can effectively encode one of random amino acid sequences synthesized in a protein 0th-order structure or 10 amino acids encoded by SNS code, are explained as follows.

1. Two amino acid sequences encoded by GC-rich $(SNS)_n$ gene and by GC-$(SNS)_n$-NSF(a), are quite different from each other, also similarly to the case of GC-rich $(GNC)_n$ gene and GC-$(GNC)_n$-NSF(a), respectively, because one of amino acid sequences, which were produced by random joining of 10 amino acids at a ratio of SNS code, is encoded on GC-$(SNS)_n$-NSF(a).
2. In addition, immature proteins, which were produced from GC-$(SNS)_n$-NSF(a), should have a smaller hydrophobicity causing more flexible structure necessary to adjust amino acid residues on the protein to a newly encountered organic compound (Fig. 8.6, Chap. 7: Fig. 7.5), although the catalytic activity would be quite low.

Those are also similar to the case of GC-$(GNC)_n$-NSF(a). Thus, a new ds-$(SNS)_n$ gene encoding a new mature protein was effectively generated through maturation of the immature protein encoded by GC-$(SNS)_n$-NSF(a).

3. In addition, degeneracy, G or C, at the third codon position enlarged the range of choice of 10 amino acids encoded by SNS code on a GC-$(SNS)_n$-NSF(a) for

efficient creation of an EntNew gene/protein without change of amino acid sequence encoded by an original gene.

3.3.4 Creation of an Entirely New Gene Under the Universal Genetic Code (GC-NSF(a) Hypothesis)

Mechanism generating an EntNew gene/protein under the universal genetic code is also essentially the same with that under the GNC code shown in Fig. 8.5. Therefore, the formation process of EntNew gene under the universal genetic code can be understood by replacing $(GNC)_n$ gene and $(GNC)_n$-NSF(a) with GC-rich gene and GC-NSF(a), respectively, except the first step of "creation of random ds-$(GNC)_n$ sequence" (Fig. 8.5).

Every modern mature protein also must be created from an immature protein-2 (Chap. 3: Fig. 3.11), which was produced by expression of a nonstop frame on antisense strand of a GC-rich gene (GC-NSF(a)) (Fig. 8.5) as described below, because it is impossible to directly create any mature protein through random processes (Chap. 3, Sect. 3.1) (Ikehara 2014). (1) An immature protein-2 was produced by expression of GC-NSF(a) (Fig. 8.5) (Ikehara et al. 1996). Note that the GC-NSF(a) also encodes one of random codon sequences, which is selected out from a protein 0th-order structure, because the GC-NSF(a) can encode a polypeptide, which satisfies the six conditions (the four main conditions plus acidic amino acid content and basic amino acid content) for protein structure formation (Ikehara et al. 1996). Successively, a mature protein could be created by accumulation of amino acid replacements of the immature protein-2, which are necessary to raise the catalytic activity as memorizing codon sequences encoding effective amino acid replacements, which were induced by random base substitutions onto the GC-NSF(a) (Fig. 8.6).

Here, I explain the maturation process from an immature gene/protein to a mature gene/protein under the universal (standard) genetic code (Fig. 8.5).

(1) First, a GC-rich gene-(1) is duplicated into two genes to conserve an original genetic function on one of the duplicated genes.
(2) An immature protein is produced by expression of the GC-NSF(a)-(1) (Fig. 8.5).
(3) Many possible catalytic sites would appear on the surface of the immature flexible protein, because the number of combinations should reach to several hundreds and because swinging of the surface residues on the immature protein further increases the number of combinations generating effective catalytic sites (Chap. 3: Sect. 4.7).
(4) If one catalytic function can be detected from many candidates for an organic compound on the immature protein-(1), the catalytic activity could be elevated to a level of sufficiently high activity by receiving multiple base substitutions necessary to raise the activity (Fig. 8.5).

(5) In parallel, structure of the immature protein gradually increases rigidity to refine the catalytic activity and to be able to reject any other organic compounds from the active site.
(6) Simultaneously, the original mature gene-(1) on sense strand and the GC-NSF(a)-(1) on antisense strand become a new immature GC-NSF(a) gene-(2) and a new mature gene-(2), respectively (Fig. 8.5).

Note that EntNew gene could not be generated from antisence sequence of AT-rich gene, because many AT (or AU)-rich stop codons (UAA, UAG and UGA) appear on the antisence sequence and, therefore, peptides produced by expression of the AT-rich antisence sequence are only short fragments. In addition, imaginary polypeptides produced by expression of antisense sequence of AT-rich gene could not be folded into water-soluble globular structure because the imaginary polypeptides cannot satisfy the six conditions for protein synthesis (Ikehara et al. 2002).

On the other hand, it would be natural that an EntNew gene could be produced from GC-NSF(a) at a high probability, because GC-NSF(a) has been derived from GC-$(GNC)_n$-NSF(a) and GC-$(SNS)_n$-NSF(a), from both of which EntNew gene could be generated at a high probability. It is important to understand that all catalytic activities, which are necessary for extant organisms to live, could be found somewhere on surfaces of a number of immature globular proteins, which are produced from GC-NSF(a)s, although the activities at the initial evolutionary stage might be, of course, extremely low. All extant organisms are living on this planet using EntNew genes/proteins and their homologous progeny genes/proteins, which were derived from the respective first family genes/proteins. Inversely stating that, modern organisms could not emerge on the present Earth, if the mechanism, which generates the first EntNew family gene/protein and homologous genes/proteins in the family, which were derived from the first family gene/protein, could not be acquired.

Further, the reasons, why an EntNew gene is effectively created from a GC-NSF(a), are as follows (Ikehara et al. 1996).

(1) Nonstop frame appears on antisense strand of GC-rich gene or GC-NSF(a) at a high probability, because three termination codons, UAA, UAG and UGA, are AU-rich.
(2) Two amino acid sequences of proteins encoded by GC-rich gene and its GC-NSF(a) are quite different from each other, because two strands of DNA are antiparallel and the universal genetic code is degenerated at third codon position (Fig. 8.8).
(3) Furthermore, the degeneracy at the third codon position of GC-rich gene could also contribute to generate various amino acid sequences encoded by GC-NSF(a) without replacing amino acid sequence of a protein encoded by the gene on sense strand (Fig. 8.8). Thus, the degeneracy could also assist the creation of an EntNew gene/protein (Ikehara 2016a).
(4) Polypeptide chain produced from GC-NSF(a) under the universal genetic code are folded into water-soluble globular structure at a high probability, because

A Q P E T G K	Amino acid sequence
5′GCNCAYCCNGARACNGGNAAR 3′	Partial GC-rich gene
3′CGNGURGGNCUYUGNCCNUUY 5′	Partial GC-NSF(a)
Y M W F C S F R V R L R P L S (R) S T G G G A	Amino acid sequences of possible imaginary proteins encoded by GC-NSF(a)

Fig. 8.8 Amino acid sequences encoded by GC-rich gene and the corresponding codon sequence on GC-NSF(a). The two amino acid sequences are quite different from each other. Possible amino acid sequences of imaginary proteins encoded by GC-NSF(a)s are also shown. Note that the amino acid sequence encoded by the sense sequence remains unchanged upon base substitutions at the third codon position of the GC-rich gene owing to degeneracy at the third codon position, although amino acid sequences encoded by the GC-NSF(a) can be largely changed (Ikehara 2016a)

the polypeptide chain satisfies well the six conditions for water-soluble globular protein formation (Ikehara 2002; Ikehara et al. 2002).

(5) A mature protein encoded by an extant gene is generally rigid (Berg et al. 2002). On the contrary, an immature protein produced from a GC-NSF(a) should be more flexible than a mature protein, because of a small hydrophobicity and a high Gly content of the protein (Ikehara et al. 1996). The high flexibility of the immature protein gives a clue for creation of a mature protein from the immature protein (Fig. 8.6).

3.4 Evidence Showing that Modern Entirely New Proteins Have Been Produced Through Essentially Random Process

3.4.1 Appearance Frequency of Neighboring Two Amino Acids in Proteins

According to my idea about the origin of protein-II, it is expected that every mature protein including every homologous protein is ultimately formed from an immature protein-2 with an essentially random amino acid sequence (Fig. 8.6). If the idea is correct, the appearance frequency of neighboring two amino acids in modern bacterial proteins should be coincident with that of neighboring the two amino acids expected from multiplication of the two average amino acid contents of the bacterial proteins. It was found that four hundreds of dots (20 (amino acids) times 20 (amino acids)) were located around the line with slope 1 in the both cases of *Helicobacter pyroli* proteins and *Haemophilus influenzae* proteins, as was expected (Ikehara 2002, 2016b). The results clearly indicate that the microbial proteins have originated from immature proteins-2 with essentially random amino acid sequence, which is arranged by random codon sequences on GC-NSF(a)s. Simultaneously, it was found that proteins used in the two bacteria carrying rather AT-rich genome were inherited from the ancestor immature proteins with random amino acid sequences encoded by GC-NSF(a)s.

3.4.2 Dependency of Hydrophilic Amino Acid Content of a Modern Protein on the Amino Acid Number of the Protein

As well known, hydrophilic amino acids and hydrophobic amino acids locate on the surface and in the core region of a water-soluble globular protein at a high probability, respectively. On the other hand, the surface area and the internal volume expand proportionally to the square and the cube of the radius of a sphere, respectively. Therefore, it is expected that the number of hydrophilic amino acids become smaller, as the number of amino acids composing a protein become larger, if the whole protein structure is consisted of a single domain. On the contrary, the ratio of hydrophilic amino acids used in a protein should be constant independently of the number of amino acids of the protein, if the protein is created by random joining amino acids, which are selected out from within a specific amino acid composition or protein 0th-order structure.

From the results analyzed with *Escerichia coli* proteins, it was found that hydrophilic amino acid content in proteins encoded by *E. coli* genome was independent of the number of amino acids in the proteins. The results clearly indicates that *E. coli* proteins have been created by essentially random polymerization of amino acids selected from in the same amino acid composition, or protein 0th-order structure, as was expected (Ikehara 2002, 2016b). Similar results were also obtained with *H. inflienzae* and *Caulobecter crescentus*. These also mean that large proteins consisted of more than about 200 amino acids must be formed with plural number of structural units or domains. We confirmed that such large proteins encoded by *E. coli* genome are generally composed of plural number of domains using about 120 amino acids.

Based on the properties of a GC-NSF(a) as described above, it can be concluded that a GC-NSF(a) effectively encodes one of amino acid sequences, which are produced by random joining of amino acids in the protein 0th-order structure, and, therefore, an EntNew mature water-soluble globular protein is created through maturation of an immature protein with essentially random amino acid sequence, which was expressed from a GC-NSF(a), at a high probability.

3.4.3 Direct Evidence for Creation of a Mature Protein from an Immature Protein-2

In order to get a more direct evidence for the GC-NSF(a) hypothesis assuming that an EntNew protein has been created from a GC-NSF(a), every base sequence encoded by GC-rich *P. aeruginosa* PAO1 genome (GC content = 66.6%) was transformed into antisense sequence (GC-NSF(a)) (Oi and Ikehara 2018). Amino acid sequence of every imaginary protein encoded by the GC-NSF(a) was homology-searched against all extant proteins encoded by the same bacterial genome. From the results, it was found that a partial amino acid sequence (63 bases long) encoded by the GC-NSF(a) of transaldolase B (*tal)* gene has a sufficiently high homology with a partial amino acid sequence of cell division protein, FtsZ, encoded by *ftsZ* gene. In addition, another result showing that a part of amino acid sequence encoded by GC-NSF(a) of major facilitater super family transporter gene encoded by the *P. aeruginosa* genome is sufficiently homologous with that of extant ABC transporter ATP-binding protein was also obtained. These mean that EntNew proteins have been generated from the respective immature proteins-2 encoded by GC-NSF(a)s carried by GC-rich genes.

Extant microorganisms have evolved from the first life for about 4 billion years and reached a sufficiently high evolutionary level so that almost any EntNew protein would not be required at this time point. In addition, homology between amino acid sequence of a mature protein, which originated and evolved from an immature protein, and amino acid sequence of an original immature protein-2 encoded by a GC-NSF(a), should be rapidly lost (Oi and Ikehara 2018). Therefore, it is crucially significant that a meaningful homology between even a partial amino acid sequence of the extant FtsZ protein and that of the imaginary protein encoded by GC-NSF(a) of *tal* gene was detected in the situations.

3.4.4 Pan-GC-NSF(a) Hypothesis on the Origin of Gene-II

As explained in Sects. 3.3.1, 3.3.3 and 3.3.4, the mechanism generating an EntNew gene/protein is quite similar under the three genetic codes or GNC genetic code, SNS genetic code and the modern genetic code. Therefore, pan-GC-NSF(a) hypothesis is named to the three hypotheses together, GC-$(GNC)_n$-NSF(a) hypothesis, GC-$(SNS)_n$-NSF(a) hypothesis and GC-NSF(a) hypothesis, as simply for convenience (Fig. 8.9). Similarly, pan-GC-NSF(a) is named GC-$(GNC)_n$-NSF(a), GC-$(SNS)_n$-NSF(a) and GC-NSF(a) together.

Strengths of pan-GC-NSF(a) Hypothesis According to the necessary conditions for the origin of something, it could be concluded that pan-GC-NSF(a) hypothesis is valid as the origin of gene-II, because creation process of an EntNew gene could be well explained as a start point of an immature pan-GC-NSF(a) encoding an immature protein with essentially random amino acid sequence, and also because the evolutionary process from the first $(GNC)_n$ gene to modern genes can be well explained (Fig. 8.9).

Weaknesses of Pan-GC-NSF(a) Hypothesis I have considered that I could explain the origin of gene-II with pan-GC-NSF(a) hypothesis. On the contrary, the origin of gene-II could not be explained by theories and hypotheses proposed by other researchers. However, it does not always mean that the creation process of EntNew

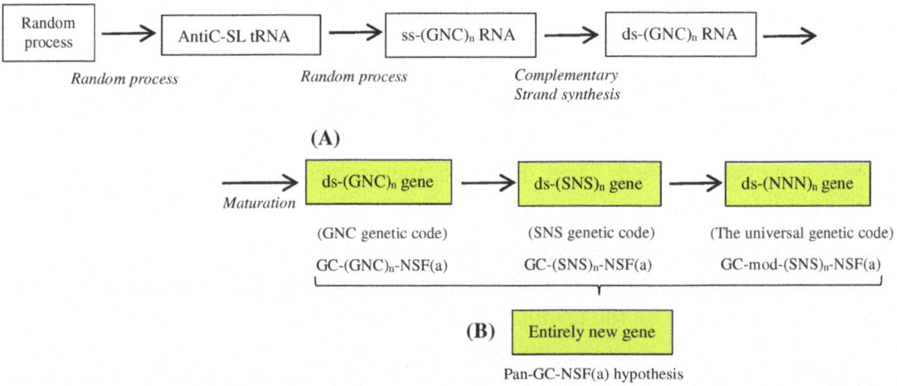

Fig. 8.9 The origin and evolutionary process of gene. (**a**) Creation process of the first ds-$(GNC)_n$ gene encoding a mature [GADV]-protein. Acquisition of AntiC-SL tRNA triggered the formation of the first ds-$(GNC)_n$ RNA gene, as suggested by AntiC-SL tRNA hypothesis. (**b**) Creation mechanism of entirely new genes. The mechanism is fundamentally the same among in three genetic code eras, because EntNew genes could be created from the respective nonstop frames on antisense strands (NSF(a)) of the corresponding GC-rich genes or GC-$(GNC)_n$-NSF(a), GC-$(SNS)_n$-NSF(a) and GC-NSF(a), which is a modified GC-$(SNS)_n$-NSF(a) sequence (GC-NSF(a) = GC-mod-$(SNS)_n$-NSF(a)), as expected by pan-GC-NSF(a) hypothesis. The hypothesis indicates that EntNew genes were and have been created through essentially random process.

genes deduced by the pan-GC-NSF(a) hypothesis is correct, based on that the creation process could be reasonably explained by the pan-GC-NSF(a) hypothesis. Similarly, it does not always mean that EntNew genes were created as deduced from the pan-GC-NSF(a) hypothesis, based on that the creation process could not be reasonably explained by other ideas.

Therefore, needless to state, it is quite important to confirm with experiments whether the creation process of EntNew genes deduced by the pan-GC-NSF(a) hypothesis is correct or not. Thus, the greatest weakness of the pan-GC-NSF(a) hypothesis is also that the creation process of EntNew genes have not been confirmed by experiments. For example,

1. Could an EntNew gene be really created from a pan-GC-NSF(a) as deduced from the pan-GC-NSF(a) hypothesis? and so on.

4 Discussion

The study on the origin of gene-I is naturally to explore how the first gene was created through random process on the primitive Earth. On the other hand, there is the fact that many EntNew genes have been created after the creation of the first gene. Therefore, in this chapter, I first described that there are two sides in the studies on the origin of gene. (1) The first one is the problem, at which the studies on the origin of gene-1 usually aim in order to answer the question, "How was the first gene encoding a mature protein created on the primitive Earth?" (2) The second one is the problem on the mechanism how and from where an EntNew gene, which does not show any meaningful homology with every previously existed gene and, therefore, does not belong to any gene family, has been created when necessary, after the first $(GNC)_n$ gene was created. Then, I discuss the problems on the origin of gene as dividing it into the two.

The Origin of Gene-I: How Was the First Gene Created?
I have proposed the anticodon joining hypothesis for explaining the way how the first gene was created on the primitive Earth about 4 billion years ago.

4.1 Anticodon Joining Hypothesis

I have considered with confidence that the first ss-$(GNC)_n$ RNA was created by random joining of anticodons carried by four nonspecific AntiC-SL tRNAs. However, several important questions have remained unanswered as follows.

(1) What generated asymmetry between two AntiC-SL tRNAs bound vertically? In other words, why could anticodons of AntiC-SL tRNAs juxtaposed at one side be joined to generate genetic information?
(2) As can be seen in Fig. 8.3, it is expected that two anticodons could be joined easily. However, as a matter of course, at least several tens of codons are usually arranged in a gene. How so many anticodons of the nonspecific AntiC-SL tRNAs could be joined by phosphodiester bond?

It is necessary and important to answer the questions in a future to complete the AntiC-SL tRNA hypothesis.

4.1.1 Qualitative Change from RNA without Genetic Information to RNA Gene

Qualitative change occurred during evolution from a RNA with random GNC codon sequence to a RNA gene. The first ss-$(GNC)_n$ RNA was naturally a simple chemical compound encoding only a random [GADV]-amino acid sequence, because RNA with an ordered sequence for production of a mature protein never be produced during random joining of GNC anticodons/codons, and because only an immature [GADV]-protein-2 could be produced from the first ss-$(GNC)_n$ RNA. Successively, the first ds-$(GNC)_n$ RNA was produced by complementary strand synthesis of the first ss-$(GNC)_n$ RNA using as a template. Nevertheless, only random $(GNC)_n$ codon sequences were arranged on both strands of the first ds-RNA. However, acquisition of the first ds-$(GNC)_n$ RNA made it possible to create the first $(GNC)_n$ RNA gene encoding a mature [GADV]-protein through maturation from an immature

[GADV]-protein. The evolutionary process from an immature protein to a mature protein was lead by promotion of catalytic activity of the immature protein (see also Chap. 3: Sect. 4.6). Thus, qualitative change occurred from a mere RNA without genetic information to RNA with genetic information for synthesis of a mature protein accompanied by maturation of an immature protein to a mature protein.

In parallel, as well known, synthesis of a mature protein has been always controlled by genetic function, after creation of the first ds-$(GNC)_n$ RNA gene. Thus, the leading role has been apparently reversed from protein to gene, after the creation of the first ds-$(GNC)_n$ RNA gene encoding a mature [GADV]-protein (Chap. 3: Sect. 4.6).

The Origin of Gene-II: How Were Entirely New Genes Created from the Past to the Present?

It would be one of the important problems in the field of biological sciences, how and from where an EntNew gene has been created. However, the important problem has remained unsolved still now. The reason is because there are several matters, which make it difficult to solve the problem, as described below.

(1) Genetic information for protein synthesis is written as one-dimensional base or actually codon sequence. On the contrary, catalytic function is expressed on a surface of a water-soluble globular protein with three-dimensional structure.
(2) On the other hand, such an EntNew gene must be also created through random process. As described several times in this book, a contemporary mature protein, which specifically recognizes one substrate, is produced through gene expression under the modern genetic system. However, it would be impossible to create a gene encoding such a mature protein with an ordered sequence at one stroke through a random process, because of a quite high wall of extraordinary sequence diversity ($\sim 10^{180}$) for creation of EntNew gene (Chap. 3: Fig. 3.4).
(3) It would be also impossible to create an EntNew gene independently of a protein, because an EntNew gene for the protein, which is totally different from any previously existed protein, cannot be designed in advance. Of course, any mature protein with a specific amino acid sequence also cannot be created independently of gene, because the gene is directly connected with amino acid sequence of the protein.

Taking the contradictory requests for creation of a mature EntNew gene/protein into consideration, the key holding for solving the difficult problem or only one possible way would be to start from an immature protein produced by random polymerization of amino acids in a specific amino acid composition or protein 0th-order structure, which is encoded by one of three GC-rich codon sequences or GC-$(GNC)_n$-NSF(a), GC-$(SNS)_n$-NSF(a) and GC-NSF(a), because the immature water-soluble globular protein could evolve to a mature protein through base replacements of the antisense sequence, using memory capacity of ds-RNA or ds-DNA.

After the creation of the first ds-$(GNC)_n$ gene, the immature proteins-2, which were produced under the protein 0th-order structure from antisense strand of the GC-rich $(GNC)_n$ gene, have triggered to create the most ancestral mature protein of a protein family. That is, GC-$(GNC)_n$-NSF(a) could overcome the quite difficult problem, how an EntNew gene could be produced, by using protein 0th-order structure (Fig. 8.5). GC-$(SNS)_n$-NSF(a) under SNS primitive genetic code succeeded the excellent properties acquired by GC-$(GNC)_n$-NSF(a) (Fig. 8.5). Even every modern mature protein must be created from an immature protein-2, which was produced by protein synthesis from a GC-NSF(a), because it is impossible to directly create any mature protein through random process in any era.

4.2 Grounds for Pan-GC-NSF(a) Hypothesis

It is the unquestionable fact that a large number of homologous genes/proteins in a gene/protein family have been newly formed according to the "gene-duplication theory". However, it is also true that the process, how EntNew genes/proteins could be created, cannot be explained by the theory. Therefore, it is obvious that the second problems of the origin of gene-II have not been solved as yet. In these case too, protein 0th-order structure holds the key for solving the riddle of the second problem of the origin of gene-II.

Then, enumerate grounds for pan-GC-NSF(a) hypothesis based on the protein 0th-order structure.

(1) Base compositions at three codon positions of modern GC-NSF(a) is comparatively similar to $(SNS)_n$, which encodes water-soluble globular protein at a high probability under the SNS code (Chap. 7: Fig. 7.4) (Ikehara 2002; Ikehara et al. 2002) (Sect. 3.3.2).
(2) Frequency using G-start codons in modern genes is sufficiently higher than those of C-, A- and U-start codons, in the cases of not only GC-rich genes but also rather AT-rich genes, indicating that [GADV+E]-amino acids constitute a basis of all proteins and, therefore, are used in proteins at a high frequency, irrespective of GC content of a gene (Ikehara et al. 1996, 2002). This also supports the idea that modern genes are successors of the most ancestral $(GNC)_n$ gene (Fig. 8.10).
(3) As shown in Fig. 8.10b, average GC content of $(GNC)_n$ genes used under the GNC primeval genetic code, was quite high as about 83%. GC content of $(SNS)_n$ genes originated from the GC-rich $(GNC)_n$ genes was also high as about 83%. After establishment of SNN primitive code through degeneracy at the third codon position, new codons were further incorporated into SNN code in order of A-start and U-start codons (Fig. 8.10b) (Ikehara 2019). Consequently, GC content of genes used under the universal genetic code has expanded from about 25% to 75%. However, GC content of codon sequences on antisense strands, which can be used for creation of EntNew genes, has remained still now in the GC-rich region expanding from about 60% to 70% as remnants of extremely GC-rich $(SNS)_n$ (~83%) and $(GNC)_n$ (~83%) genes (Fig. 8.10b) (Ikehara et al. 1996, 2002).

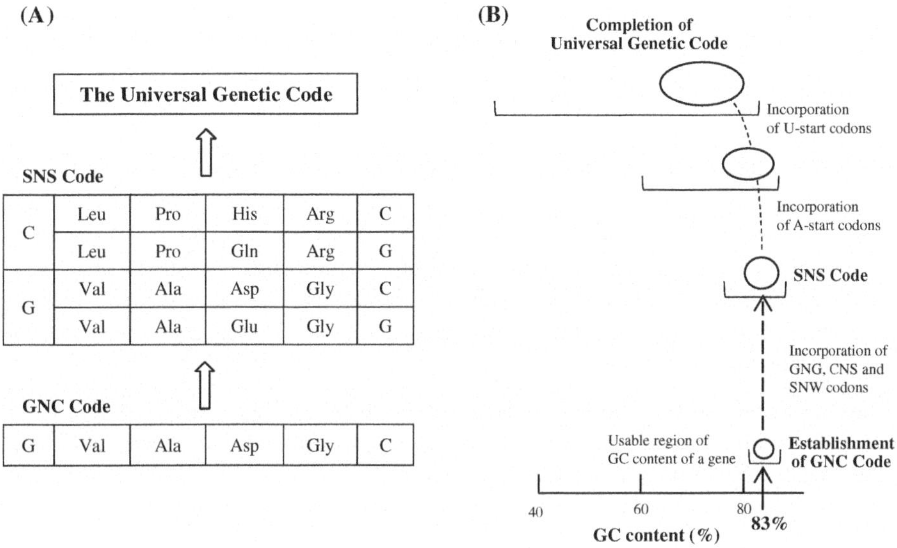

Fig. 8.10 (**a**) Evolutionary pathway of the genetic code. The evolutionary process is drawn according to GNC-SNS primitive genetic code hypothesis, assuming that the modern genetic code originated from GNC code encoding four [GADV]-amino acids (Ikehara et al. 2002). (**b**) Change of the field for EntNew gene creation, accompanied by the genetic code evolution. After establishment of GNC code, codons were incorporated into the GNC code in order of GNG, CNS, GNW, CNW, A-start and U-start codons (Ikehara 2019). Upon the evolution of the genetic code, the field for EntNew gene creation shifted from extremely GC-rich (around 83%) to ordinary GC-rich region around 60 to 70%. Circles and ellipsoids indicate the supposed fields for EntNew gene creation. Upward brackets show the supposed usable region of GC content of a gene under the respective genetic codes

(4) In GNC primeval genetic code era, maturation from an immature gene to an EntNew gene was chiefly adjusted by exchanging Gly for Ala encoded by GGC and GCC, respectively (Chap. 7: Fig. 7.6). On the contrary, in SNS code era, hydrophilic amino acids encoded by SAS codons were predominantly exchanged for hydrophobic amino acids encoded by SUS codons on the evolutionary process from an immature protein to a mature protein (Fig. 8.7).
(5) On the other hand, evolutionary process from an immature protein encoded by GC-NSF(a) to a mature protein under the universal genetic code has been controlled by the two adjustments, one is the exchange of Gly for Ala and the other is replacement of hydrophilic amino acids with hydrophobic amino acids. These also indicate that modern GC-rich genes and GC-NSF(a)s have been propagated from the most ancestral $(GNC)_n$ gene through $(SNS)_n$ gene.

On the contrary, EntNew gene/protein cannot be produced from antisense sequence of AT-rich gene (Fig. 8.10b). The reasons are explained in brief as follows.

(1) Stop codons appear on the antisense strand of AT-rich gene at a high frequency, because all the three termination codons, UAA, UAG and UGA, are AU- or AT-

rich. This means that only short peptides should be synthesized from the antisense codon sequences of AT-rich genes.

(2) Imaginary protein encoded by codon sequence on antisense strand of AT-rich gene should be highly hydrophobic, making the protein insoluble in water (Ikehara et al. 2002).
(3) The imaginary proteins encoded by antisense codon sequence of AT-rich gene should have excess β-sheet and extremely low turn/coil formabilities and too low acidic-AA composition (Ikehara 2002). These indicate that an imaginary protein, which is produced from codon sequence on antisense strand of AT-rich gene, does not satisfy the six conditions, indicating that the imaginary protein could not form a water-soluble globular structure as a prerequisite for formation of proteinaceous catalysts.

As described so far, EntNew genes/proteins were and have been created from GC-rich codon sequence on antisense strand, irrespective of any genetic code era, GNC code, SNS code and the universal genetic code. The following properties of the genetic code enabled the efficient creation of EntNew gene/proteins.

(1) Hydrophobic amino acids and hydrophilic amino acids are arranged in U_2 and A_2 columns of the universal genetic code, respectively.
(2) In U_2 column, β-sheet (Val, Ile, Phe) and α-helix (Leu, Met) forming hydrophobic amino acids are arranged. On the other hand, turn/coil (Asn, Asp) and α-helix (Glu, Gln, Lys) forming hydrophilic amino acids are arranged in A_2 column. The excellent arrangements of the amino acids in the universal genetic code table have enabled to produce proteins with appropriate secondary structure formabilities, irrespective of GC content of a gene.
(3) Protein 0th-order structures, amino acid compositions composed of amino acids encoded by GNC code and SNS code, have written in one and four rows of the universal genetic code table, respectively.

Such a stunning arrangement of amino acids in the universal genetic code table would be the results, that appropriately new amino acids were captured and introduced during evolution of the genetic code (Chap. 7: Sect. 3.3). The universal genetic code, which has been formed through evolution from the first GNC genetic code through SNS code (Ikehara et al. 2002), made it possible to create EntNew genes/proteins efficiently and for diverse organisms to prosper on the present Earth. I have marveled anew that the beautiful life system has been acquired during the evolution.

5 Conclusion

The essence of gene is simply represented as a triplet codon sequence for synthesis of a mature protein. Why is the triplet code used in modern gene? The reason is usually considered as because 20 amino acids can be used in the triplet code. However, it would be impossible to generate the formally triplet GNC genetic code in preparation for the situation that 20 amino acids can be used in a future. The reason would be actually because GNC code was accidentally used as the first genetic code (Chap. 6: Sect. 3.3). Similarly, the reason, why the genetic code is degenerate, can be understood not only because the universal genetic code composed of 20 amino acids and 64 codons originated from the primeval triplet GNC code composed of four [GADV]-amino acids and four codons, but also because EntNew genes could be efficiently created from pan-GC-NSF(a)s except GC-$(GNC)_n$-NSF(a) owing to the degeneracy at the third codon position (Fig. 8.8).

The reason, why such a frame of the genetic system has been established, is because the first ds-$(GNC)_n$ gene was created by using four AntiC-SL tRNAs carrying GNC anticodon as receiving support of catalytic activities of [GADV]-proteins. The first ds-$(GNC)_n$ gene was created by the three key polymers, [GADV]-protein, AntiC-SL tRNA and ss-$(GNC)_n$ RNA, all of which could be produced through random processes under protein 0th-order structure or the amino acid composition of [GADV]-amino acids.

As described above, gene is a carrier of genetic information, which is written as codon sequence for synthesis of an amino acid sequence of a mature protein, which is necessary to live, into ds-RNA or ds-DNA. Then, why the genetic information is written into double-stranded RNA or DNA? The reason is because the first genetic sequence was created with two dimers, which were arranged two AntiC-SL tRNA vertically (Chap. 6: Fig. 6.9). Consequently, this would make it possible to self-replicate genetic information through the ds-RNA or ds-DNA.

New genes have been created through two pathways. One is for homologous new genes and the other is for EntNew genes. The steps for creation of homologous new genes are as follows.

1. The creation of a progeny homologous gene always requires a parental gene. Therefore, the parental gene must be first duplicated.
2. One gene is preserved to retain the original genetic information. The other is used for creation new homologous gene through introduction of base substitutions onto the previous genetic sequence.
3. Amino acid replacements are introduced though base substitutions of one duplicated gene to generate flexibility around the original active site. The generation of a flexible structure makes it possible to easily accept a new organic compound as a substrate.
4. The new active site could be formed by introduction of additional base substitutions to fit to the new organic compound.
5. Thus, it would be inevitable to path through a flexible structure in order to create a new homologous protein from a codon sequence on sense strand of a parental gene.

The steps for creation of an EntNew gene are as follows.

1. Synthesis of an immature protein with some flexibility from pan-GC-NSF(a) (Fig. 8.9).

2. Maturation of the immature protein to a new mature protein encoded by a new mature gene.

Thus, both homologous genes and EntNew genes were and have been created from a codon sequence encoding a loosen mature protein and from a pan-GC-NSF(a) encoding an immature protein with some flexibility, respectively. Thus, a mature gene could not be created directly, because any mature protein never be produced through a random process at one stroke.

References

Berg JM, Tymoczko JL, Stryer L (2002) Biochemistry, 5th edn. WH Freeman Comp, New York

Gilbert W (1986) Origin of life: the RNA world. Nature 319:618

Gilbert W (1987) The exon theory of genes. Cold Spring Harb Symp Quant Biol 52:901–905

Ikehara K (2002) Origins of gene, genetic code, protein and life: comprehensive view of life systems from a GNC-SNS primitive genetic code hypothesis. J Biosci 27:165–186

Ikehara K (2014) Protein ordered sequences are formed by random joining of amino acids in protein 0th -order structure, followed by evolutionary process. Orig Life Evol Biosph 44:279–281

Ikehara K (2016a) Degeneracy of the genetic code has played an important role in evolution of organisms. SOJ Genet Sci 3:1–3

Ikehara K (2016b) GADV hypothesis on the origin of life -Life emerged in this way!? LAP LAMBERT Academic Publishing, Saarbrucken

Ikehara K (2019) The origin of tRNA deduced from *Pseudomonas aeruginosa* 5′anticodon-stem sequence -anticodon-stem loop hypothesis. Orig Life Evol Biosph 49:61–75

Ikehara K, Yoshida S (1998) SNS hypothesis on the origin of the genetic code. Viva Orig 26:301–310

Ikehara K, Amada F, Yoshida S et al (1996) A possible origin of newly-born bacterial genes: significance of GC-rich nonstop frame on antisense strand. Nucleic Acids Res. 24:4249–4255

Ikehara K, Omori Y, Arai R et al (2002) A novel theory on the origin of the genetic code: a GNC-SNS hypothesis. J Mol Evol 54:530–538

Ohno S (1970) Evolution by gene duplication. Springer, Heiderberg

Oi R, Ikehara K (2018) Direct evidence for GC-NSF(a) hypothesis on creation of entirely new gene/protein. Curr Proteom 3:13–23

Romero MLR, Rabin A, Tawfik DS (2016) Functional proteins from short peptides: Dayhoff's Hypothesis Turns 50. Angew Chem Int Ed 55:15966–15971

Chapter 9
The Origin of Life

Abstract The riddle of the origin of life remains unsolved irrespective of strenuous efforts by many researchers. What is the major difficulty, which has become the obstacle against solving the origin of life? The reason would be as follows. All kinds of extant organisms on the present Earth are living under the core life system composed of three biopolymers; DNA/RNA carrying genetic information, tRNA realizing genetic code and protein with catalytic function. Nowadays, all the biopolymers with ordered sequence are synthesized under the genetic system, as DNA is synthesized by self-replication, both tRNA and protein are produced by expression of the respective genes. However, all the first three biopolymers, as a matter of course, must be synthesized through random processes, because the sequence of every biopolymer with ordered sequence never be designed in advance. Therefore, the greatest obstacle, which makes it difficult to solve the riddle of the origin of life, would be to explain the way how synthesis of mere polymers with random sequence could be converted to production of biopolymers with ordered sequence during the repeated random processes. For the purpose, several hypotheses including RNA world hypothesis have been proposed by other researchers. However, the origin of life has been unfortunately unsolved still now. Then, first, those hypotheses proposed thus far have been introduced and discussed to clarify the problems encompassed in the hypotheses in the first part of this chapter. Thereafter, I introduce my idea how biopolymers with ordered sequence, as protein, tRNA and gene, could be formed through random processes on the primitive Earth. Recently, I proposed anticodon stem loop hypothesis on the origin of tRNA, suggesting that the first tRNA could be produced through random processes as a small and stable hairpin loop RNA composed of only 17 nucleotides including anticodon in the loop, based on analysis of 5′ anticodon stem sequences of *Pseudomonas aeruginosa* PAO1 tRNAs. Taking it into consideration together with protein 0th-order structure for producing an immature water-soluble globular protein by random joining of [GADV]-amino acids and with the first single-stranded RNA created by random joining of GNC anticodons carried by the most primitive tRNAs, it was understood that the "three keys" for formation of the respective three biopolymers with ordered sequence could be obtained. In other words, it has become possible to explain main steps to the emergence of life with the three keys for the first time. The establishment process of the fundamental life system composed of the six members containing three

K. Ikehara, *Towards Revealing the Origin of Life*,
https://doi.org/10.1007/978-3-030-71087-3_9

other members or cell structure, metabolism and genetic code, in addition to the three keys or protein, tRNA and gene, is explained to give a solution to the serious problem of the origin of life.

Keywords Origin of life · RNA world hypothesis · Hydrothermal vent theory · Protein world theory · Amyloid world theory · α-helical peptide theory · Coenzyme world hypothesis · tRNA core hypothesis · Space-origin hypothesis · [GADV]-protein world hypothesis (GADV hypothesis)

From "Origin Of Life: Twentieth Century Landmarks: The Oparin-Haldane Hypothesis"

> Oparin suggested that the organic compounds could have undergone a series of reactions leading to more and more complex molecules. He proposed that the molecules formed colloid aggregates, or 'coacervates', in an aqueous environment. The coacervates were able to absorb and assimilate organic compounds from the environment in a way reminiscent of metabolism. They would have taken part in evolutionary processes, eventually leading to the first lifeforms. (From homepage of Chris Gordon-Smith 2003)

I am surprised at reading the description about Oparin's idea on the origin of life, which he already considered in the early years of the twentieth century. Oparin well understood how life emerged on Earth, although the detail concepts were different from GADV hypothesis described in this book (Fig. 9.1). His idea can be interpreted by rewriting with my idea as follows.

> The organic compounds ([GADV]-amino acids) could have undergone a series of reactions leading to more and more complex molecules ([GADV]-proteins). He proposed that the molecules ([GADV]-proteins) formed colloid aggregates, or 'coacervates', ([GADV]-microspheres) in an aqueous environment. The coacervates ([GADV]-microspheres) were able to absorb and assimilate organic compounds (as glyoxylate, glyceraldehyde, pyruvate

Fig. 9.1 Diverse organisms inhabit on the present Earth like as cosmoses shown in the above photograph. How could such a beautiful flower be borne on this planet as one of descendants from the first life, which emerged on the primitive Earth about 4 billion years ago? In this chapter, it is explained how the riddle of the origin of life could be solved

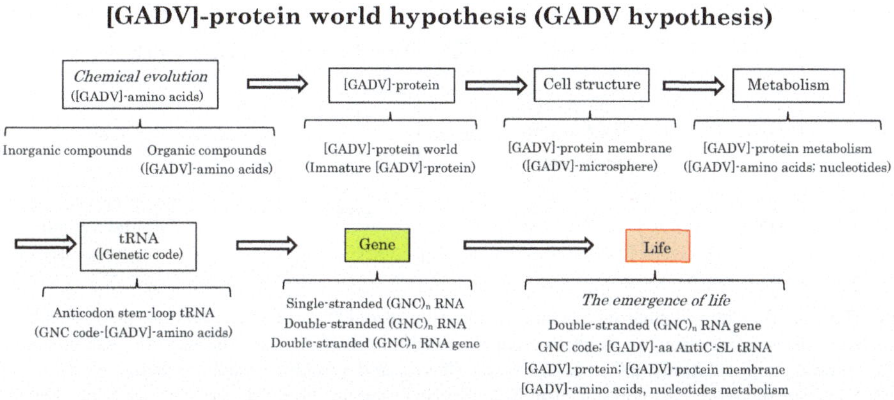

Fig. 9.2 In this chapter, the origin of life is discussed (red box). Life (red box) emerged on the primitive Earth as a result of establishment of the fundamental life system, which is composed of the six members or protein, cell structure, metabolism, tRNA, genetic code and gene. Outline of the steps from chemical evolution for syntheses of organic compounds to the emergence of life, which are deduced from [GADV]-protein world hypothesis or GADV hypothesis, is shown in the figure. It is observed that all the steps to the emergence of life are supported by [GADV]-amino acids and the related organic compounds as [GADV]-peptides, GNC genetic code encoding [GADV]-amino acids and a $(GNC)_n$ RNA gene encoding a [GADV]-protein and so on

and so on) from the environment in a way reminiscent of metabolism (driven by [GADV]-proteins). They would have taken part in evolutionary processes, eventually leading to the first lifeforms.

1 Introduction: Towards Solving the Riddle of the Origin of Life

First, it is described what matters are necessary to explore and solve the riddle of the origin of life, as similarly to the other chapters so far. Based on the matters, it is chiefly discussed how modern wonderful lives as cosmoses on the present Earth have evolved from the first life, which emerged on the primitive Earth (Figs. 9.1 and 9.2).

1.1 Necessary Matters for Exploration of the Origin of Life

1. It is necessary to well understood what features modern life has, in order to understand from what kind of primitive life the contemporary life originated.
2. If it must be assumed that the contemporary life originated from a primitive one with features different from the contemporary life, as a matter of course, the conversion process from the primitive one to the first one similar to the modern one must be reasonably explained with continuity, either directly or indirectly.

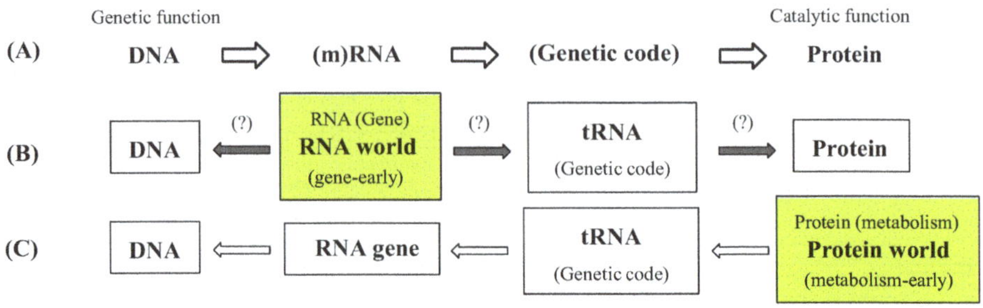

Fig. 9.3 (**a**) Genetic information flow in the direction from DNA (genetic function) or (m)RNA to protein (catalytic function) in modern organisms (white open arrows). The genetic system holds the "chicken-egg relationship" between DNA (RNA) and protein, because genetic function on DNA cannot be expressed without protein and protein cannot be produced without gene (DNA/RNA). (**b**) In RNA world hypothesis, it is considered that formation of the first fundamental life system began from RNA (RNA world) with dual functions (genetic function and catalytic function), and later, genetic and catalytic functions were transferred to DNA and protein, respectively. (**c**) In GADV hypothesis, the formation of the core life system began from protein (protein world) and successively progressed to form genetic code, RNA gene and DNA gene as going up from the lower (protein world) to the upper stream (gene) of the genetic flow in the present life system. Closed arrows and question marks mean the steps unidentified well still now

For example, in the RNA world hypothesis, it is considered that life emerged from RNA world, which was formed by RNA self-replication. In this case, it must be explained how the core life system composed of three main members was established in the RNA world comprised of self-replicated RNAs (Fig. 9.3a). Similarly, in the GADV hypothesis, it is considered that life emerged from [GADV]-protein world, which was formed by pseudo-replication of [GADV]-protein. Therefore, it must be explained how the fundamental life system was established in [GADV]-protein world comprised of pseudo-replicated [GADV]-proteins.

It could be judged that the hypothesis is invalid, by whether or not the establishment process of the fundamental life system, which is necessary to solve the origin of life, could be reasonably explained by the hypothesis.

1.2 General Structural Features of Modern Life

1. Life is a substance, which is surrounded by cell membrane to be able to discriminate from surroundings.
2. Life is composed of six members of the fundamental life system or protein, cell structure, metabolism, tRNA, genetic code and gene. Above all, the three members, protein, tRNA and gene, form the core of the system. The system composed of the three main members is called as the core life system in this book (Chap. 2: Fig. 2.1a).

1.3 General Functional Features of Modern Life

1. Life is living under the fundamental life system.
2. Life self-replicates and proliferates.
3. Life is an evolvable substance.

The riddle of the emergence of life could be understood as the result, if the establishment process of the fundamental life system composed of the six members were reasonably elucidated. On the other hand, the origins of the six members are intimately related to each other. Therefore, some figures, which should be shown in this chapter, are omitted from this chapter if possible, in order to avoid duplicate use of the figures. Instead, the place of the figure, which is given in the other chapters, is described in the text of this chapter.

2 The Ideas About the Origin of Life Advocated by Other Researchers

2.1 RNA World Hypothesis

For many years, the well-known "chicken and egg relationship" between gene and protein became a problem for solving the origin of life (Fig. 9.3a). Francis Crick, Leslie Orgel and Carl Woese speculated that a solution may involve RNA, and the discovery of ribozymes (Kruger et al. 1982; Guerrier-Takada et al. 1983) provided a foundation for the idea. It has been expected that both the "chicken-egg relationship" and the riddle of the origin of life might be solved by RNA world hypothesis, which was proposed by Gilbert (1986). The term RNA World was invented to describe an era, in which the first form of life emerged as incorporating ribozymes having both genetic and catalytic functions. It is also considered that the first genetic system later has been established by transferring genetic sequence information on the ribozyme to DNA and then the catalytic functions of the ribozymes to proteins (Fig. 9.3b). Szostak and his group developed an elegant experimental procedure called SELEX: Systematic Evolution of Ligands by Exponential enrichment to discover aptamers (Ellington and Szostak 1990: Tuerk and Gold 1990) which can bind to a variety of biochemical compounds such as nucleic acids and proteins (Thodima et al. 2006). Furthermore, Szostak et al used SELEX to produce new ribozymes from a large pool of random sequences. The generation of artificial ribozymes with RNA synthetic activity strongly supported the RNA world concept which is now described even in biochemistry textbooks.

Before the discovery of double-stranded (ds)-DNA (Watson and Crick 1953), proteins, which are the primary constituent of cells, had been considered as genetic materials. However, with the discovery of ds-DNA structure by Watson and Crick (1953), it was made clear that all organisms on the present Earth are living in "gene/

replicator-centered era" under the core life system composed of gene, tRNA (genetic code) and protein. Then, many researchers have naturally considered that gene or replicator must be first formed for the first life to emerge. At the same time, it became recognized that it is essential to elucidate formation process of the life system in order to solve the riddle of the origin of life. When the RNA world hypothesis was proposed, the consensus was that life could not emerge from proteins alone. The reason is based on two assumptions: (1) Any mature protein like a precision polymer machine cannot be produced without gene, because amino acid sequence diversity is extraordinary large – $20^{100} = \sim 10^{130}$, even for a small protein composed of 100 amino acids (Dill 1990). (2) Proteins cannot be replicated because of the absence of specific recognition between two amino acids, while RNA and DNA can selectively interact through Watson-Crick base pairs (Watson and Crick 1953). Of course, it is quite natural that many researchers has considered that proteins cannot be produced without genetic information, because modern proteins, which composed of 20 amino acids and are used in extant organisms, have evolved into like as precise polymer machines. Thus, most researchers concluded that life could not emerge from proteins alone.

Strengths of RNA World Hypothesis As well known, RNA world hypothesis on the origin of life was proposed as a chance of the discovery of ribozymes, which might have both genetic function and catalytic function. One of the reasons why the RNA world hypothesis has been widely accepted by many researchers would be because they have implicitly considered that protein never be produced without gene. Therefore, the RNA world hypothesis, which is based on gene and self-replication of the gene, has been accepted by many researchers and the hypothesis is in main stream of the study on the origin of life still now.

Weaknesses of RNA World Hypothesis However, there are many weaknesses in the RNA world hypothesis, which many researchers rely on solving the origin of life, as described below (Table 9.1). Nevertheless, many researchers may consider that it could be overcome irrespective of the difficulties, owing to the long time on the primitive Earth.

1. It is difficult to synthesize nucleotides and RNA, which are indispensable to form RNA world, with prebiotic means.
2. It would be also difficult to amplify RNA by self-replication.
3. 64 codons must be prepared before the emergence of life, because all combinations of triplet bases would appear on RNA strand produced by random process.
4. In addition, RNA, which was produced by random joining of nucleotides, could not acquire genetic information for a mature protein synthesis in the absence of protein in principle (Chap. 3: Sect. 4.8.2), even if RNA could be produced before protein was generated.

Table 9.1 Weaknesses of RNA world hypothesis

1. It is difficult to synthesize nucleotides with prebiotic means.
2. It is also difficult to synthesize RNA with prebiotic means.
3. It is difficult for RNA to self-replicate.
4. 64 codons would appear on RNA strand produced by random process.
5. It is difficult to explain how the first genetic code was established.
6. It is probably impossible for RNA to acquire genetic information.

However, the misunderstanding, that life must emerge in accordance with the gene/replicator-first theory, has made it difficult further for many researchers to solve the riddle (Ikehara 2018). On the other hand, many researchers have also gradually understood that the RNA world hypothesis might not be realized because RNA is too complex to prebiotically produce, and because RNA produced by random joining of mononucleotides could not acquire genetic information for protein synthesis (Table 9.1) (Ikehara 2017).

I have indicated that there are many weaknesses in the RNA world hypothesis (Ikehara 2017), as described in Table 9.1. I further explain here the major defects of RNA world hypothesis, with which it would probably make it impossible to overcome the defects in principle. Nevertheless, why has the RNA world hypothesis based on "gene/replicator-first" theory been accepted by many researchers up to the present time? The reason would be because they do not sufficiently recognize the significance of protein 0th-order structure or a specific amino acid composition of [GADV]-amino acids. Inversely, it cannot be considered for me that after gene was first produced, tRNA and genetic code was formed and eventually protein was synthesized under the translation system.

As described in detail in Chap. 3, gene does not play a leading role in creation of an entirely new (EntNew) mature protein from an immature protein, because the random base substitutions of ds-RNA/DNA cannot determine the direction of evolutionary process of the immature protein, but the evolutionary pathway of the protein is actually determined by functional improvement of the evolving protein (refer Chap. 3: Sect. 4.8.2). Therefore, it would be obvious that it is impossible for RNA to acquire genetic information for protein synthesis and to synthesize a mature protein according to base sequence of the RNA synthesized previously. In facts, it has been pointed out that studies from "ground zero" will be necessary to solve the riddle of the origin of life (Luisi 2014, 2016).

2.2 Hydrothermal Vent Theory

The research based on the hydrothermal vent theory targets at the place, where life might emerge and what kinds of organic compounds could be produced at the place (Corliss et al. 1979). The mineral-rich chimneys with alkaline and acidic fluids, providing a source of energy that facilitate chemical reactions between hydrogen and carbon dioxide to form increasingly complex organic compounds. In addition, it is also confirmed that the hot, alkaline conditions found at the vents successfully created protocells (Zhang et al. 2017) – regarded as a vital basic building block for life (Holm and Andersson 2005). However, the origin of life could not be solved through the research, which is not directly related to clarify the establishment process of the fundamental life system.

Strengths of Hydrothermal Vent Theory The hydrothermal vent theory is interesting in the sense that the following possibilities are provided by the theory. (1) Energy, which facilitates chemical reactions to produce organic compounds, can be supplied from the hydrothermal vent. (2) The hot, alkaline conditions at the vents promote creation of protocells (Zhang et al. 2017). Both of them are indispensable to form the first life.

Weaknesses of Hydrothermal Vent Theory However, it would be impossible to make clear the origin of life, because it could not clarify the establishment processes of the core life system composed of gene, tRNA (genetic code) and protein.

2.3 Protein World Theory

Some protein world theories on the origin of life have been presented by other researchers as described below. However, GADV hypothesis ([GADV]-protein world hypothesis), which I have proposed (Fig. 9.3c), is not discussed in this section, because the hypothesis is described in detail in the following Sect. 3, although GADV hypothesis is of course one of protein world theories on the origin of life.

2.3.1 Amyloid World Theory

Amyloid world theory is introduced in Maury's review article (2018) as follows. The amyloid world hypothesis, which was first proposed by Maury (2009), posits that in the pre-RNA world era, information processing was based on catalytic amyloids. The self-assembly of short peptides into β-sheet amyloid conformers leads to extraordinary structural stability and novel multifunctionality that cannot be achieved by the corresponding nonaggregated peptides. The new functions include self-replication, catalytic activities, and information transfer. The environmentally sensitive template-assisted replication cycles generate a variety of amyloid polymorphs on which evolutive forces can act, and the fibrillar assemblies can serve as scaffolds for the amyloids themselves and for ribonucleotides, proteins and lipids.

Strengths of Amyloid World Theory One of the characteristics of life is self-replication. The amyloid theory has been proposed as focusing on the way how the first life acquired the ability of self-replication. It is confirmed by experiments that β-sheet amyloid conformers, which were formed by assembly of short peptides, self-replicated. It is also supposed that generation of a variety of amyloid polymorphs could lead evolution of the amyloids, which have abilities for self-replication, catalytic activities and information transfer.

Weaknesses of Amyloid World Theory However, the origin of life could not be solved through the research, because the self-replication ability of β-amyloid cannot be transferred to ds-RNA/DNA and the establishment process of the core life system or gene and tRNA could not be explained based on the β-amyloid world theory.

2.3.2 α-Helical Peptide Theory

Ghadiri 's group succeeded to obtain evidence showing that two half-sized α-helical peptides could be joined on another full length of α-helical peptide as a template. This is quite interesting in the meaning that protein or peptides also could be self-replicated (Lee et al. 1996; Li and Chmielewski 2003).

Strengths of α-Helical Peptide Theory It is significant to show a more generalized system that protein also can be replicated without any involvement of genetic function.

Weaknesses of α-Helical Peptide Theory However, it would be quite difficult to produce one long and two short peptides, which were used in the protein self-replication experiments, with prebiotic means in the absence of gene. Therefore, it is doubtful whether such a self-replication system of peptides could be created on the primitive Earth. Furthermore, it would be also difficult that the self-replication system composed of coiled-coil structures contribute to solution of the riddle of the origin of life, because it would be impossible to transfer the self-replication system with α-helices to one carried out with ds-RNA/DNA.

2.4 Coenzyme World Hypothesis

Coenzymes have been also presented as candidates of organic compounds, which can be self-replicated (Sharov 2009, 2016). Primordial life had no nucleic acids; instead heritable signs were represented by isolated catalytically active self-reproducing molecules, similar to extant coenzymes, which presumably colonized surfaces of oil droplets in water. Further he assumes that coenzyme-like molecules

(CLMs) changed surface properties of oil droplets (e.g., by oxidizing terminal carbons), and in this way created and sustained favorable conditions for their own self-reproduction.

Further, it is described in Sharov's paper (2016) as follows. Polymerization of CLMs became controlled by other polymers used as templates; and this kind of template-based synthesis eventually resulted in the emergence of RNA-like replicons. Apparently, oil droplets transformed into the outer membrane of cells via engulfing water, stabilization of the surface, and osmoregulation. In result, the metabolism was internalized allowing cells to accumulate free-floating resources (e.g., amino acids, ATP), which was a necessary condition for the development of protein synthesis. Thus, life originated from simple but already functional molecules, and its gradual evolution towards higher complexity was driven by cooperation and natural selection.

Strengths of Coenzyme World Hypothesis The coenzyme world hypothesis has been proposed as focusing on the way how the first life acquired the ability of self-replication. In the hypothesis, the steps to the emergence of life are explained based on self-replication through catalytic activities of coenzyme-like molecules, cell structure composed of oil droplets, self-reproduction, protocell formation, protein synthesis and so on.

Weaknesses of Coenzyme World Hypothesis However, it should be the most important point for solving the origin of life, how the first genetic system was established. Nevertheless, the steps to the emergence of life are not concretely explained. For example, through what catalytic steps with coenzyme-like molecules nucleotides and RNA molecule could be synthesized, and how the RNA synthesized with coenzyme-like molecules could be acquired genetic information for a mature protein synthesis, and by what the template-based synthesis could be carried out to generate RNA-like replicons, and so on.

2.5 tRNA Core Hypothesis

De Farias et al. (2016) have recently proposed "tRNA core hypothesis" for solving the origin of life. The central role of tRNA molecules in the origin and evolution of the fundamental biological processes is emphasized. For example, they consider that tRNAs, which were at the core of a prototranslation system, gave origin to the first gene (mRNA), the peptidyl transferase center (rRNA), proto-tRNAs, the anticodon and operational codes. The authors also advocate that metabolic pathways emerged from evolutionary pressures of the decoding systems. The transitions from the RNA world to the ribonucleoprotein world to modern biological systems were driven by three kinds of tRNAs transitions, to wit, tRNAs leading to both mRNA and rRNA.

Strengths of tRNA Core Hypothesis It is certainly interesting in that the authors have payed attention to a wide range of roles of tRNA in especially bridging over between genetic information and amino acid sequence of a protein, because there is a high probability that the riddle of the origin of life could be clarified through exploration of tRNA. In fact, as described above, the authors have recognized that tRNA occupies a key position connecting among the first gene (mRNA) and the peptidyl transferase center (rRNA), proto-tRNA, which should be at the core of a prototranslation system.

Weaknesses of tRNA Core Hypothesis However, the steps to the emergence of life are not unfortunately concretely explained, for example, the processes, how genetic code for translation could be established and how genetic information for protein synthesis could be formed, are not indicated.

2.6 Space-Origin Hypothesis

The researches who value the space-origin hypothesis on the origin of life chiefly target at ancestral organisms and organic compounds in or from space, of which discoveries might contribute to the emergence of life. However, the research could not clarify the riddle of the origin of life through the establishment process of the fundamental life system, until a newly emerged life can be discovered in space, in meteorites or in interstellar dusts from space.

For example, it was described in newspapers as "a journey for exploring for the origin of life", when Hayabusa 2 was launched on 3 December, 2014. however, it is now described only that "Hayabusa 2 will target a C-type asteroid "Ryugu" to study the origin and evolution of the solar system as well as materials for life by leveraging the experience acquired from the Hayabusa mission. Furthermore, we expect to clarify the origin of life by analyzing samples acquired from a primordial celestial body such as a C-type asteroid to study organic matter and water in the solar system and how they coexist while affecting each other" in the article published recently on website of JAXA (May 22, 2020), Japan. However, I have to state that it would be probably impossible to clarify the establishment process of the fundamental life system and, therefore, to solve the riddle of the origin of life, even if organic compounds could be detected from samples of the "Ryugu", which have been acquired by the Hayabusa 2-mission.

Recently, Furukawa et al. (2019) reported that ribose and other sugars were detected from primitive meteorites including Murchison meteorite. However, I consider that the detection of ribose from the meteorites does not support RNA world hypothesis at all, because ribose is still an intermediate for nucleotide synthesis and, in addition, the steps to the emergence of life could not be explained from the standpoint of the RNA world hypothesis, even if nucleotides could be synthesized with ribose in space.

Strengths of Space-Origin Hypothesis With flourishing of recent astrobiology, researchers have actively studied in accordance with the space-origin theory on the origin of life. From the results, useful knowledge as extraterrestrial organic synthesis has been obtained. Furthermore, it may become possible to understand necessary conditions for the emergence of life widely more than ever before, if extraterrestrial life could be found.

Weaknesses of Space-Origin Hypothesis Comparative studies of organisms living on the Earth with extraterrestrial life could be carried out, if the extraterrestrial lives were discovered. However, the fundamental problem on the origin of life never be solved even by the discovery of the extraterrestrial lives, because the discovery never lead to elucidation of the establishment process of the fundamental life system.

2.7 Common Weaknesses of Other Hypotheses for Exploration of the Origin of Life

Thus, as described in the above Sects. 2.1, 2.2, 2.3, 2.4, 2.5 and 2.6, the ideas based on only a few viewpoints for grasping the essence of life, such as self-replication (heredity), catalytic reaction (metabolism), tRNA (prototraslation system) and so on, have been proposed thus far (Table 9.2). The respective viewpoints on the origin of life, which were presented in the ideas by other researchers, are of course significant and interesting. However, in such cases also, it is necessary to show not individually but comprehensively the way how the fundamental life system comprising of the six members was established and the first life emerged. However, I do not know any idea at the present time, under which the steps from chemical evolution to the emergence of life could be well explained (Table 9.2).

Therefore, it would be unfortunately impossible to explain the origin of life with the respective ideas because of the causes as described below.

1. Establishment process of the fundamental life system, which is the most important point to clarify how the first ife emerged through random process on the primitive Earth, has not been well discussed and explained in any other hypotheses, probably because they paid too much attention to a few points of the fundamental life system or of features of modern life (Table 9.2).
2. Therefore, transition process from the respective chemical materials to the first life living under the fundamental life system could not be explained.

Table 9.2 Main remarkable points of research carried out under other hypotheses on

Hypothesis	Remarkable points	
RNA world hypothesis	Self-replication	Catalytic reaction
Hydrothermal vent hypothesis	Catalytic reaction	Protocell formation
Amyloid world hypothesis	Self-replication	Catalytic reaction
Cofactor world hypothesis	Catalytic reaction	Protocell formation
tRNA core hypothesis	Prototranslation system	Mediator between gene and protein

3 My Idea About the Origin of Life

3.1 [GADV]-Protein World Hypothesis (GADV Hypothesis)

Elucidation of the origin of life is to clarify the establishment process of the fundamental life system composed of the six members as already described in the respective chapters. Therefore, many descriptions in this chapter may sometimes overlap the contents described in the respective chapters. Even the cases, please read this chapter and consider again GADV hypothesis as a review because the origin of life is the main theme of this book.

Then, I explain the establishment process of the fundamental life system leading to solution of the riddle of the origin of life, based on the following main points.

1. It is described as attaching weight to the establishment process of three main members of the core life system, as protein, tRNA (genetic code) and gene (Chapter 2: Fig. 2.1a).
2. It is essential to make clear the steps, how the three members were formed through random processes, only which occurred on the primitive Earth.
3. It is important to clarify the origins of the three members under the same concept, not independently but comprehensively.
4. It is also important to explain evolutionary process from the most primitive three members to the respective modern ones.

Then, I first explain how I could reach the GADV hypothesis (Ikehara 2002, 2005) as an alternative of RNA world hypothesis (Ikehara 2017). About fifteen years ago, I proposed the GADV hypothesis which I reached mainly based on computer analysis of databases of extant microbial genes and proteins. The hypothesis assumes that life emerged from a [GADV]-protein world, which was formed by pseudo-replication of [GADV]-protein in the absence of any genetic function (Ikehara 2009). I reasoned that [GADV]-protein should be pseudo-replicated if immature water-soluble globular [GADV]-proteins with a weak but sufficiently high activity for peptide bond formation could be produced on the primitive Earth. Such a protein could then synthesize other water-soluble immature [GADV]-proteins having a similar globular structure by direct random joining of [GADV]-amino acids without genetic guidance (see also Chap. 3: Sects. 4.2, 4.3 and 4.4).

At first, it would be difficult to produce proteins composed of one hundred of [GADV]-amino acids in prebiotic conditions (the number of 100 is proposed here, because a small globular protein or a single domain protein is usually composed of around 100 amino acid residues) (Dill 1990). Therefore, [GADV]-protein would actually be aggregates of [GADV]-peptides, which assembled by non-covalent forces such as hydrophobic interactions and ionic bonds (positive ions as divalent metal ions could be supplied from circumstances). Water-soluble globular structures could be formed by the aggregation of peptides, if they were composed of roughly equal amounts of [GADV]-amino acids. Based on computer analysis of data of extant microbial genes encoding water-soluble globular proteins (Ikehara et al. 2002), it could be supposed that peptide fragments composed of the four amino acids should be incorporated into secondary structures such as α-helices, β-sheets and turn/coils, and tertiary structure as their aggregates would be stabilized by hydrophobic interaction among the fragments. In other words, the earliest protein-like molecules evolved from aggregates of short [GADV]-peptides to [GADV]-protein composed of about 100 amino acid residues. For simplicity in the rest of this chapter, I name the water-soluble peptide aggregate as [GADV]-protein. We have experimentally confirmed that [GADV]-proteins in the form of aggregated [GADV]-peptides, can be synthesized by repeated wet-drying procedures of [GADV]-amino acids. Furthermore, the polymers should have exhibited not only protease activity but also peptide bond forming activity as the reverse reaction (Oba et al. 2005). Other researchers have also confirmed that di- or tri-peptides exhibit catalytic activity (Jacobsen et al. 1999; Shibasaki et al. 2004) and that even peptides as simple as Gly-Gly, and Gly-Gly-Gly can catalyze peptide bond formation between chemically activated amino acids (Gorlero et al. 2009). The [GADV]-protein world hypothesis is based on the sequence of events in which the first genetic system was established to synthesize increasing amounts of [GADV]-proteins as evolving from globular aggregates of short [GADV]-rich peptides to more complete water-soluble globular proteins. I will briefly describe the process by which I arrived at the set of four [GADV]-amino acids as components of primitive water-soluble globular proteins. The research was begun with an analysis of seven microbial genomes in terms of the amino acid composition related to secondary and tertiary structures (Ikehara et al. 1996). Then, the research developed to a hypothesis related to a genetic code that involved ten amino acids encoded by the primitive SNS genetic code (Ikehara and Yoshida 1988; Ikehara et al. 2002), and finally to the GADV hypothesis presented here (Chap. 7: Sect. 3.2). The weak but sufficiently high catalytic activity of the protein led to the emergence of life. Therefore, the hypothesis is an idea based on the "metabolism (protein)-early theory" (Ikehara 2018). Thus, I consider that the first genetic system was established to synthesize more and more refined [GADV]-protein, from globular aggregate of short [GADV]-rich peptides to more complete water-soluble globular [GADV]-protein with a high molecular weight, step by step.

3.2 Establishment Process of the Fundamental Life System

Here, consider how the core life system composed of three main members or gene, tRNA and protein was established from which member to verify the validity of the GADV hypothesis (Fig. 9.4).

As described above, a mature protein like a precision polymer machine never be produced through random process without the corresponding gene. Therefore, mature proteins are always formed under the control of genes in these days. On the other hand, a mature gene also cannot be created directly through random process (Chap. 3: Sect. 4.3). Then, how was the fundamental life system established? For the purpose, three formation processes are considered *a priori*.

Pathway 1: Gene-first theory: Many researchers implicitly consider that gene must be first produced, because it is generally considered that protein cannot be produced in the absence of gene. Therefore, the establishment process of the core life system is generally considered according to the Pathway 1 that gene carrying genetic information for protein synthesis was first created. After that, tRNA was created to express the genetic information of the gene so that protein could be produced. That is, it is considered by many researchers that the life system was established along the genetic flow. However, gene never be formed without tRNA and protein (Chap. 7: Sect. 3.2).

Pathway 2: tRNA-first theory: In this case, it is considered that both gene and protein were produced after tRNA connecting gene with protein was first created, for example, as expected by tRNA core hypothesis on the origin of life (De Farias et al. 2016). However, the first tRNA never be produced through random process independently of catalytic activity of protein (Chap. 6: Figs. 6.9 and 6.11). Furthermore, it would be obvious that genetic code also never be established without tRNA, which is intimately related to formation of genetic code.

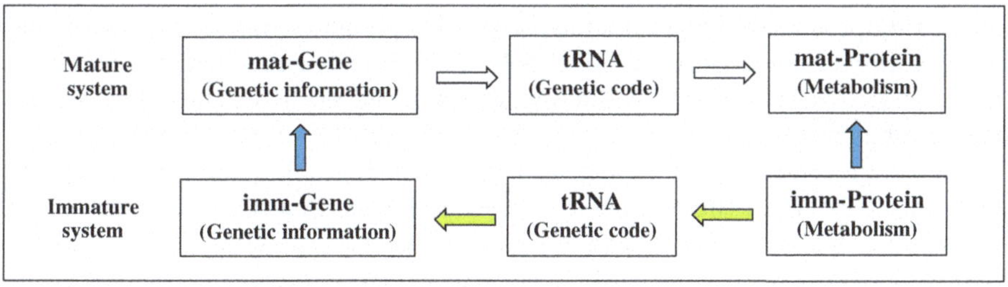

Fig. 9.4 Under the modern fundamental life system, mature (mat) proteins are always produced under expression of genes with genetic information for synthesizing the corresponding protein (white bold arrows). On the contrary, the fundamental life system was established in order of immature (imm) protein, primeval tRNA and immature gene (yellow bold arrows). After establishment of the immature system, the system evolved to the mature life system, as shown with blue bold arrows

Inversely stating this, it means that the first tRNA should be generated with the catalytic activities of immature water-soluble globular [GADV]-proteins, which could be produced by random joining of [GADV]-amino acids (Chap. 6: Sect. 3.4). Therefore, it is assumed that life emerged as a trigger of formation of water-soluble globular but immature [GADV]-protein with some flexibility produced through random process (Fig. 9.4).

Pathway 3: Protein-first theory: It is well known that a mature protein like a precision molecular machine never be generated through random process. Therefore, the pathway 3 assumes that not a mature but an immature water-soluble globular protein was first created, thereafter tRNA was created using catalytic functions of the immature proteins, and finally, the first gene encoding a mature protein was created through evolutionary process from the immature protein to a mature protein as using the memorizing ability of ds-RNA, which was produced by immature [GADV]-proteins (Chap. 3: Sect. 4.6). In the pathway 3, it is considered that the core life system was established as going to upstream against the genetic flow. Therefore, only one possible way for creation of a mature protein is as follows. Immature but water-soluble globular proteins were first produced by random joining of [GADV]-amino acids and successively tRNA was produced by the immature [GADV]-proteins.

4 Three Keys for Solving the Riddle of the Origin of Life

As described up to here, it is considered that life emerged at the time point, when all the six members constituting the fundamental life system have been created through the respective random processes. The establishment order of the fundamental life system are shown in Fig. 9.5. Above all, the formation processes of the three members or [GADV]-protein, AntiC-SL tRNA and ds-$(GNC)_n$ gene are the core for the emergence of life (Fig. 9.5). In addition, as can be seen in Fig. 9.5, it could be understood that the establishment process of the fundamental life system is always founded by protein 0th-order structure or [GADV]-amino acids. Furthermore, the fact is worthy of note that AntiC-SL tRNA, which was underestimated in the “central dogma” (Chap. 2: Fig. 2.1a), dominated the lead in the establishment of the fundamental life system.

I noticed recently that there are three keys, which are the core for solving the riddle of the origin of life and made it possible for the first life to emerge on the primitive Earth. Then, I would like to introduce the three keys and also to summarize my idea [GADV]-protein world hypothesis or GADV hypothesis as a review of the hypothesis, at the substantially last chapter of this book.

I have considered that life emerged as [GADV]-protein world hypothesis or GADV hypothesis expects, which is based on “protein/metabolism-first” theory (Ikehara 2018), because [GADV]-protein could be pseudo-replicated in the absence of gene owing to protein 0th-order structure or [GADV]-amino acids, and also

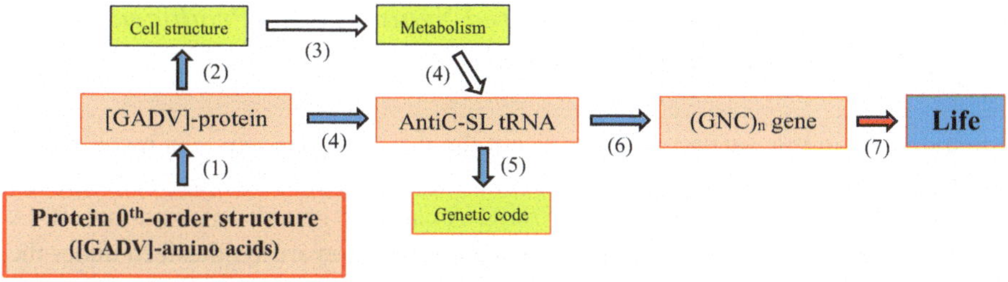

Fig. 9.5 According to the GADV hypothesis, the steps to the emergence of life can be reasonably explained as follows. (1) Immature [GADV]-proteins, actually aggregates of [GADV]-peptides, were produced by random joining of [GADV]-amino acids under protein 0th-order structure. (2) Cell structure or [GADV]-microsphere was constructed by aggregation of the [GADV]-proteins. In parallel, [GADV]-protein world was formed in the [GADV]-microsphere. (3) In the [GADV]-protein world, the initial metabolic system was formed. (4) AntiC-SL tRNA was produced by catalytic reactions with immature [GADV]-proteins in the initial metabolism. (5) The GNC primeval genetic code was established accompanied by the formation of four specific AntiC-SL tRNAs. (6) Double-stranded (ds) $(GNC)_n$ gene was formed through complementary strand synthesis of ss-$(GNC)_n$ RNA, which was produced by random joining of anticodons carried by the AntiC-SL tRNAs. (7) The first life eventually emerged upon the formation of ds-$(GNC)_n$ genes

because the replication ability is not always required from the beginning as assumed by the "gene/replicator-first theory" but it is sufficient if gene/replicator was acquired at some time point until the emergence of life (Ikehara 2018).

Therefore, some readers may consider that major difficulties were already overcome by the GADV hypothesis. However, two difficulties, which are intimately related each other, still remain unexplained in detail. One is how mere chemical materials could be transformed to biopolymers with ordered sequence, and the other is how the fundamental life system could be established on the primitive Earth during repeated random processes. I had already obtained two keys, which are required for formation of two biopolymers through random processes, one is immature water-soluble globular [GADV]-protein with some flexibility produced under protein 0th-order structure or [GADV]-amino acids in the absence of genetic function (Ikehara 2014), and the other is the first ss-$(GNC)_n$ RNA created by random joining of GNC anticodons in tRNA, which encodes random [GADV]-amino acid sequence (Ikehara 2002, 2005). In addition, I have, recently, proposed anticodon stem loop hypothesis on the origin of tRNA, suggesting that a small and stable AntiC-SL composed of only 17 nucleotides was the first tRNA, which could be produced through random processes (Chap. 6) (Ikehara 2019). This means that three keys have been obtained, which are necessary to solve the riddle of the origin of life. Then, I discuss in the following section, how the "three keys" can unlock the three doors necessary to enter the respective rooms of protein, tRNA and ss-RNA gene, all of which have ordered sequence, and finally the first life emerged owing to the three keys.

4.1 *Can Biopolymers with Ordered Sequence Be Produced Through Random Processes?*

I first discuss in this section whether or not the three biopolymers can be produced during random processes.

In modern organisms, gene or DNA/mRNA carrying genetic information, tRNA realizing genetic code, and protein with catalytic function are synthesized under the fundamental life system, all of which are composed of the respective biopolymers with ordered sequence (Fig. 9.6a). Note here that DNA or RNA itself does not generally exhibit any concrete function as like a catalyst, but only expresses genetic information indirectly through syntheses of proteins and a small number of tRNAs and rRNAs. Moreover, function of tRNA is also expressed by mediating between genetic information and amino acid sequence. In contrast, only protein, which was produced through random process, can directly exhibit its catalytic function (Fig. 9.7).

Of course, only mere chemical materials should be produced under random processes on the primitive Earth before establishment of the fundamental life system (Chap. 3: Fig. 3.5). More and more complex organic compounds and various kinds of polymers could be produced during the repeated random reactions. However, all of them should be still mere chemical materials irrelevant to life activities. Therefore, the greatest obstacle, which makes it difficult to solve the riddle of the origin of life, would be to explain the way how synthesis of the polymers with random sequence could be converted to production of biopolymers with ordered sequence, as like gene, tRNA and protein, during repeated random processes (Chap. 3: Sect. 4.6, Chap. 6: Fig. 6.11). Stating this inversely, if the conversion processes could be

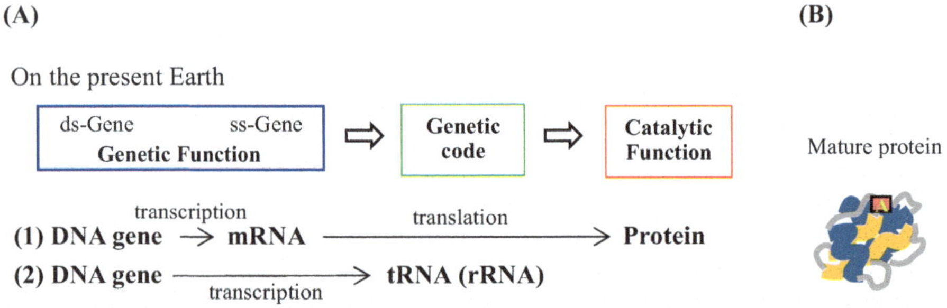

Fig. 9.6 (**a**) All polymeric members with ordered sequence, as tRNA (rRNA) and protein playing roles in translation, are produced under genetic function in the fundamental life system. Here, mRNA is regarded as single-stranded gene. All members in the life system of modern organisms, mRNA, tRNA, rRNA and protein, including DNA gene itself, are produced under the genetic function of ds-DNA. Therefore, modern organisms on the present Earth are apparently living in a “gene-centered era”. (**b**) Mature protein synthesized under the life system is generally made of one amino acid sequence, which is folded into one tertiary structure expressing one catalytic activity. Blue and yellow thick curves and gray thin curves show three secondary structures, α-helix, β-sheet and turn/coil, respectively

elucidated, the way, how the life system composed of the biopolymers with ordered sequence was established during repeated random reactions, could be made clear. This means that two difficulties described above, the conversion from mere polymers with random sequence to biopolymers with ordered sequence and the establishment process of the fundamental life system, could be solved at one stroke. However, the difficulty of the conversion process from polymers as mere materials to biopolymers with function also has confronted researchers who want to solve the riddle of the origin of life for many years. On the other hand, the fact, that various kinds of organisms inhabit on the present Earth owing to the three members with ordered sequence, clearly indicates that mere polymers with random sequence were converted to biopolymers with ordered sequence one day on the primitive Earth.

As I have already pointed out in the previous chapters that a novel concept, or protein 0th-order structure, must be introduced to overcome the difficulty of formation of a meaningful protein during random process (Chap. 3: Sects. 4.2, 4.3 and 4.4) (Ikehara 2009, 2014). Therefore, the first key is a flexible water-soluble globular protein, which is produced by random joining of [GADV]-amino acids or under the protein 0th-order structure. One of the reasons why a water-soluble globular protein could be produced even by random joining of the amino acids is because the respective amino acids themselves can function for protein structure formation. This is distinctly different from mononucleotide, which is, itself, meaningless for formation of genetic information on DNA or RNA.

In next section, I will discuss how the three members, DNA/RNA, tRNA and protein, with the respective ordered sequences could be produced during random processes in reverse order of genetic flow, protein, tRNA and gene, because only protein out of the three members could be produced by direct random joining of its monomeric units or [GADV]-amino acids on the primitive Earth (Ikehara et al. 2002; Ikehara 2005, 2009, 2014). Although the first protein with some function, which could be produced during random process, was only an immature-[GADV]-protein with some flexibility, the immature protein could trigger formation of the fundamental life system (Fig. 9.5) (Chap. 3: Sects. 4.2, 4.3 and 4.4).

4.2 *How Could the Three Keys for Solving the Riddle of the Origin of Life Be Formed During Random Processes*

4.2.1 The First Key: Immature Water-Soluble Globular [GADV]-Protein with Some Flexibility

The first key is an immature but flexible water-soluble globular [GADV]-protein or proto-[GADV]-protein, actually an aggregate of [GADV]-peptides, which could be synthesized by direct random joining of [GADV]-amino acids in a special amino acid composition containing [GADV]-amino acids at roughly equal amounts. Such a special amino acid composition is one of protein 0th-order structures (Ikehara 2014). (Chap. 3: Sect. 4.4). The reason, why a proto-[GADV]-protein could be

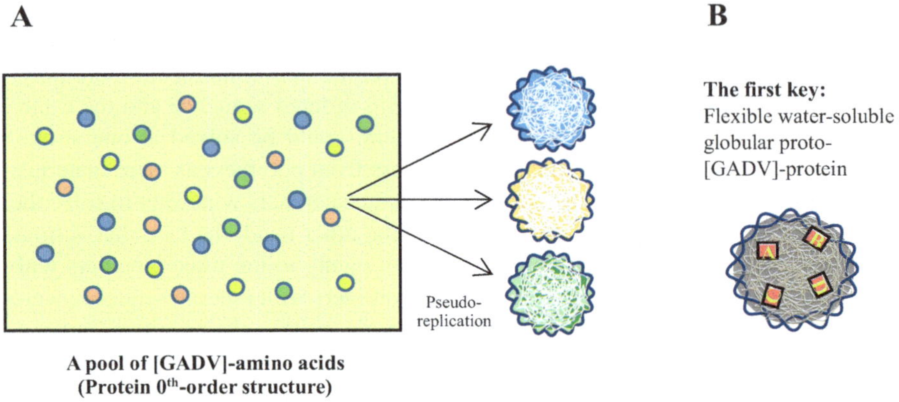

Fig. 9.7 (**a**) Water-soluble globular immature [GADV]-proteins with some flexibility could be produced by direct random joining of [GADV]-amino acids drawn by small colored circles, because amino acid composition of [GADV]-amino acids is one of protein 0th-order structures. Immature [GADV]-proteins synthesized under the protein 0th-order structure have a similar amino acid composition but quite different amino acid sequence from each other. This kind of protein synthesis was named as pseudo-replication (Ikehara 2009). The immature [GADV]-proteins, actually aggregates of [GADV]-peptides, functioned as the first key for establishment of the fundamental life system. Three similar, but different proteins produced are drawn as lightly colored circles surrounded by wavy lines. (**b**) Many low but sufficiently high catalytic centers, as shown with alphabets, A, B, C and D, could be appeared on the surface of a flexible water-soluble globular protein at a high probability (Chap. 3: Sect. 4.7). Both wavy lines around the circle and thin yellow curves in the circle symbolize flexibility of the immature [GADV]-protein

produced even by random joining of [GADV]-amino acids, is because the protoprotein satisfies the following four conditions (hydropathy, α-helix, β-sheet and turn/coil formabilities) for protein structure formation. Therefore, it is supposed that the [GADV]-polypeptide chains could be folded into water-soluble globular structure with some flexibility at a high probability (Fig. 9.7). [GADV]-protein world was formed by pseudo-replication of immature [GADV]-proteins in [GADV]-microsphere (Ikehara 2002, 2009), which were generated, for example, through repeated wet-drying cycles of [GADV]-amino acids in depressions on rocks of seaside on the primitive Earth (Oba et al. 2005; Suwannachot and Rode 1998). The first primeval metabolic system could be formed in the protein world and various chemical reactions necessary to the emergence of life could be catalyzed by the immature [GADV]-proteins (Chap. 5: Sect. 3.1).

4.2.2 The Second Key: Anticodon Stem-Loop tRNA

The second key is a small but sufficiently stable hairpin loop RNA or AntiC-SL tRNA (Ikehara 2019). The idea was hit upon from analysis of sequence similarity among 5′ anticodon-stem (5′ AntiC-stem) sequences of *P. aeruginosa* PAO1 tRNAs (tRNAdb (Universitat Leipzig)), suggesting that vestiges of the origin and evolution of tRNA are left in the 5′ stem sequences of *P. aeruginosa* tRNAs (Chap. 6: Sects. 3.2 and 3.3). Based on the results, I concluded that the first primeval tRNA was AntiC-SL. The reason is as follows (Ikehara 2019). (1) 5′ AntiC-stem sequences exhibited the closest relation among three stem-loop structures, D-stem loop (D-SL), AntiC-SL and T-SL, and one accepter stem of *P. aeruginosa* tRNA (Chap. 6: Sect. 3.2) (Ikehara 2019). (2) The evolutionary process of tRNA deduced from the neighborhood relationships of the 5′ AntiC-stem sequences was roughly coincident with the evolutionary process assumed by GNC-SNS primitive genetic code hypothesis (Ikehara et al. 2002; Ikehara 2019) (Chap. 6: Sect. 3.3).

The minimum sized hairpin-loop RNA with sufficient stability could be formed during repeated random joining of nucleotides and degradation of unstable oligonucleotide by immature [GADV]-proteins (Fig. 9.8a). In facts, it has been reported that active sites of extant nucleotide metabolizing enzymes, such as DNA-dependent and RNA-dependent RNA polymerases are composed of mainly [GADV]-amino acids (Van der Gulik et al. 2009), suggesting that primeval and immature [GADV]-proteins have catalytic activities for both RNA synthesis and RNA degradation. Thus, the second key or AntiC-SL tRNA mediating between genetic information and amino acid sequence of a protein could be produced through random processes owing to the first key or an immature [GADV]-protein. The RNA gene encoding the first AntiC-SL tRNA played an important role in generating four primeval AntiC-SL tRNAs for [GADV]-peptide synthesis. That is, the second primeval AntiC-SL tRNA could be created by synthesis of complementary strand of the first AntiC-SL tRNA (Ikehara 2019). Furthermore, the third and the fourth AntiC-SL tRNA genes could be also easily produced through duplication of the first or the second AntiC-SL tRNA gene followed by introduction of necessary base substitutions onto the AntiC-SL tRNA genes. The creation of new genes for nonspecific AntiC-SL tRNAs made it possible to form the four specific [GADV]-AntiC-SL tRNAs composed of 5 base-pairs stem and 7 bases loop carrying GNC anticodon (Chap. 6: Fig. 6.9).

These imply that the second and the subsequent AntiC-SL tRNAs could be produced by succeeding to one of previously existed AntiC-SL tRNAs. Therefore, only the first AntiC-SL tRNA was formed by random process (Fig. 9.8a). Slightly flexible and immature water-soluble globular [GADV]-proteins with one of random [GADV]-amino acid sequences could be also produced under the four AntiC-SL tRNAs for translating GNC codons to [GADV]-amino acids. In this case too, the protein 0th-order structure largely contributed to produce the immature [GADV]-proteins (Fig. 9.8b). I would like to stress again that a promising primeval tRNA, which was produced during random processes and exhibited meaningful function for translation, would be the smallest and stable AntiC-SL tRNA.

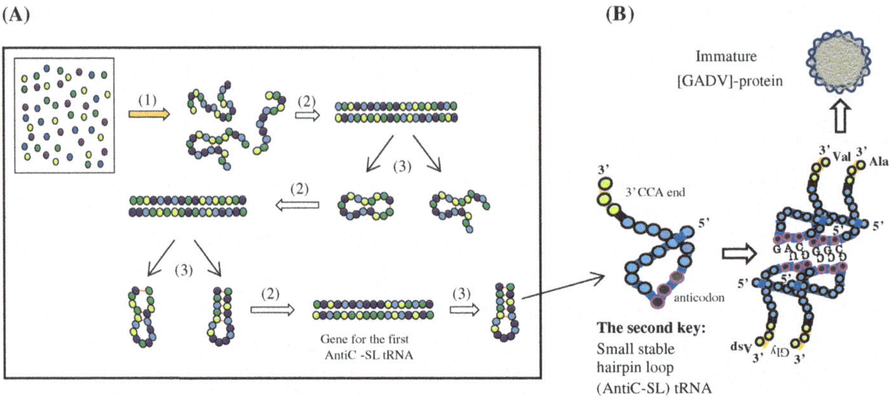

Fig. 9.8 (**a**) A small and stable hairpin loop RNA could be produced after the following reaction steps (Note that Fig. 9.8 (A) is the same with Chap. 6: Fig. 6.11). (1) Synthesis of RNA by random joining of mononucleotides shown with four small colored circles, (2) degradation of unstable RNA with a smaller number of base pairs than others, followed by complementary strand synthesis, and (3) error-prone transcription by immature [GADV]-protein. Steps (2) and (3) were repeated. A stable and the smallest hairpin loop RNA finally obtained became the first anticodon stem-loop (AntiC-SL) tRNA. (**b**) AntiC-SL tRNA played a role in random [GADV]-protein synthesis as the second key for establishment of the first GNC genetic code. The AntiC-SL tRNAs could play two roles of original tRNA and mRNA each other. However, only immature [GADV]-protein could be produced under the four kinds of primeval AntiC-SL tRNAs. The AntiC-SL additionally carrying CCA sequence was drawn by tracing AntiC-SL and the 3′-CCA end of a modern L-form yeast Phe-tRNA (PDBj; 1EHZ)

4.2.3 The Third Key: Single-Stranded $(GNC)_n$ RNA

The third key is the first ss-$(GNC)_n$ RNA, which was formed by random joining of anticodons located in the loop of four primeval AntiC-SL tRNAs for [GADV]-protein synthesis (Fig. 9.9). However, any mature protein like a precision polymer machine could not be synthesized even by the expression of the ss-$(GNC)_n$ RNA, because the ss-$(GNC)_n$ RNA was composed of only random GNC codon sequence. Only a water-soluble globular immature [GADV]-protein could be produced (Fig. 9.9b). The acquisition of the ss-$(GNC)_n$ RNA enabled to produce many immature [GADV]-proteins with the same amino acid sequence and also to form the first ds-$(GNC)_n$ RNA through complementary strand synthesis of the ss-$(GNC)_n$ RNA. In this manner, ds-$(GNC)_n$-gene encoding a mature protein was formed through maturation of the first ds-$(GNC)_n$ RNA at the last stage of the establishment process of the fundamental life system. That is, the first gene, which expresses its function at the most upstream from protein synthesis in the genetic flow, was formed at the last stage toward the emergence of life (Fig. 9.2). This is consistent with the ideas that gene is meaningless before formation of tRNA and/or protein and that life emerged as expected by not "gene/replicator-first theory" but "protein/metabolism-first theory" (Ikehara 2018).

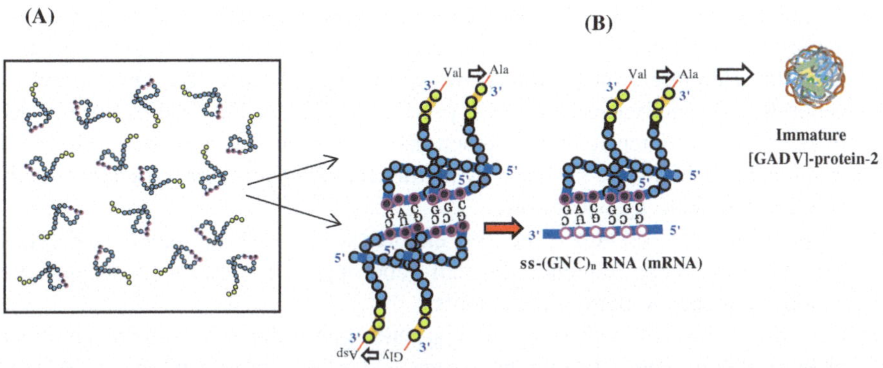

Fig. 9.9 (**a**) Four AntiC-SL tRNAs were randomly selected out from a pool of four kinds of AntiC-SL tRNAs. Side-by-side dimer formation between the two nonspecific AntiC-SL tRNAs carrying one of [GADV]-amino acids at 3′ terminal end could stimulate [GADV]-protein synthesis because of adjoining effect of two amino acids (not shown in the figure). The complexes could also form another type of dimer, one complex opposites to another complex through base pair between two complementary GNC anticodons in the AntiC-SL tRNAs. (**b**) The first single-stranded (ss)-$(GNC)_n$ RNA could be produced by joining of anticodons in AntiC-SL tRNAs randomly placed side by side, owing to catalytic activity of immature [GADV]-proteins. The ss-$(GNC)_n$ RNA made it possible to synthesize a number of [GADV]-proteins-2 with the same amino acid sequence for the first time, although the proteins were still immature

4.2.4 The Genuine Life Emerged Upon Acquisition of Double-Stranded $(GNC)_n$ Gene

The core life system composed of gene, tRNA (genetic code) and protein was established by using the three keys as going up from downstream to upstream of the genetic flow, that is, in order of immature [GADV]-protein, AntiC-SL tRNA (GNC genetic code), ss-$(GNC)_n$ RNA and the first ds-$(GNC)_n$ RNA gene. Although the ds-RNA played a quite important role in the emergence of life, the first ds-$(GNC)_n$ RNA was not the fourth key, because the formation of ds-$(GNC)_n$ RNA through complementary strand synthesis of the first ss-$(GNC)_n$ RNA is a nonrandom process.

(1) Only immature and flexible water-soluble globular proteins were produced by expression of either of two sequences of the first ds-$(GNC)_n$ RNA, which was formed just after complementary strand synthesis of the ss-$(GNC)_n$ RNA, because essentially random GNC codon sequences were simply arranged on the two strands. However, various kinds of catalytic functions could be expressed on the surface of one immature pluripotent [GADV]-protein with some flexibility (Chap. 3: Sect. 4.7), although their catalytic activities were low (Fig. 9.9b). Accordingly, if one catalytic activity necessary to live could be appeared on the immature protein, the immature protein evolved to one mature [GADV]-protein as accumulating necessary base replacements onto the first ds-$(GNC)_n$ RNA, because it was possible to preserve base substitutions on the ds-RNA for the first time.

(2) Acquisition of the first ds-$(GNC)_n$ RNA gene was essential in producing various mature [GADV]-proteins with a high catalytic activity like a precision polymer machine. Hereafter, one mature protein with a rigid structure, which exhibits generally one catalytic activity, was produced from one ds-$(GNC)_n$ RNA gene. In parallel, inefficient catalytic reactions carried out by immature [GADV]-proteins until then were replaced by mature [GADV]-proteins with a more reliable catalytic activity one by one.
(3) Acquisition of ds-$(GNC)_n$ gene also enabled to increase the number of genes with the same base sequence through gene duplication and to propagate the genetic sequence to many descendants.
(4) Gene duplication of the first ds-$(GNC)_n$ gene opened the way to create a number of homologous genes from sense strand encoding the first mature [GADV]-protein. In addition, the first ds-$(GNC)_n$ gene did also contribute to create EntNew genes from the antisense strand, which could encode an immature protein with some flexibility. The immature protein could evolve to another mature protein, if one necessary catalytic activity to live could be found on the immature protein. Thus, a large number of both homologous genes and EntNew genes could be generated from sense and antisense sequences, respectively (Sect. 3.3). In this manner, mature proteins expressing a high catalytic function could lead to the emergence of life.
(5) Formation of the first ds-$(GNC)_n$ gene opened a kind of the "gene-centered era" about 3.8~4.0 billion years ago. As described above, the three keys opened the respective doors in order of formations of immature [GADV]-protein, the first AntiC-SL tRNA and the first ss-$(GNC)_n$ gene, all of which were produced through the respective random processes (Fig. 9.10). The first life could emerge on the primitive Earth after acquisition of various kinds of ds-$(GNC)_n$ genes. After that, all organisms lived on the past Earth and are living on the present Earth in a kind of "*gene-centered era*". Organisms also will live forever in the seemingly "*gene-centered era*". The seemingly "*gene-centered era*" means that proteins substantially hold all the functions behind genes back, although all organisms have lived under the control of genes (see also Chap. 3: Sect. 4.8). The leading role in the life system was apparently reversed from protein to gene, accompanied by the acquisition of ds-$(GNC)_n$ genes encoding a mature [GADV]-protein.

4.2.5 The Steps to the Emergence of Life Viewed from the Three Keys

As described up to this point, I am now convinced that the conversion processes from inanimate materials to living matter could be made clear upon the discoveries of the three keys for understanding how the fundamental life system was established and for solving the riddle of the origin of life. Several interesting characteristics about the establishment processes of the life system can be seen in Fig. 9.10.

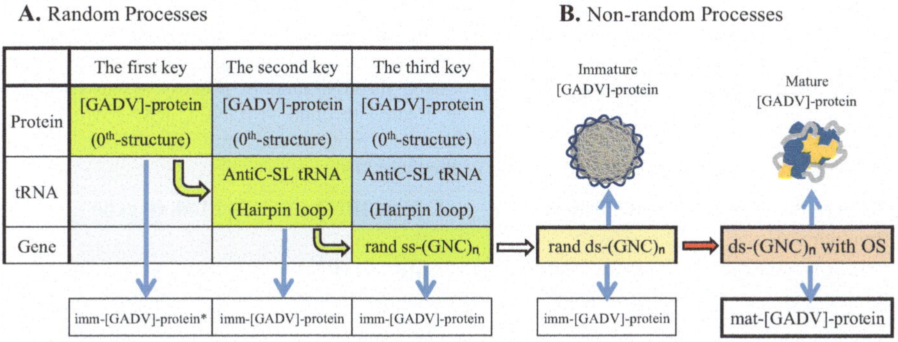

Fig. 9.10 (**a**) Three biopolymers, protein, tRNA and RNA, were produced on the primitive Earth in order of [GADV]-protein, AntiC-SL tRNA and single-stranded (ss)-$(GNC)_n$ RNA gene as going up from downstream to upstream of the genetic flow. Yellow curved thick arrows indicate that the first and second keys contributed to formation of the subsequent second and third keys, respectively. Formation of the third key was assisted by not only the second key shown in yellow box but also the first key described in pale blue box. Downward blue arrows indicate protein synthesized under the respective systems. The first imm-[GADV]-protein* (the most left, bottom box) was actually an aggregates of [GADV]-peptides containing organic compounds as simple alkyl amines and alkyl carboxylic acids. In contrast, immature (imm)-[GADV]-protein means that the protein was composed of only [GADV]-peptides with random (rand) amino acid sequence. (**b**) Mature (mat)-protein like a precision polymer machine with ordered sequence (OS) could be first produced under the ds-$(GNC)_n$ gene. White bold arrow indicates ds-$(GNC)_n$ RNA formation by complementary strand synthesis of ss-$(GNC)_n$ RNA. Red bold arrow indicates maturation process from ds-$(GNC)_n$ RNA encoding an immature protein with random amino acid sequence to ds-$(GNC)_n$ RNA gene encoding a mature protein with ordered amino acid sequence

(1) It is important to note that the protein 0th-order structure or [GADV]-amino acids always existed behind the formation of the three keys.
(2) The three keys could play the respective roles in establishing the life system in order of the first protein, the first tRNA (GNC primeval genetic code) and the first ss-$(GNC)_n$ RNA before the emergence of life.
(3) A hierarchy can be seen in the formation processes of the three keys (Table 9.3). That is, formation of the first key, *immature water-soluble globular [GADV]-protein with some flexibility*, does not require any other keys. However, formation of the second key, *the first anticodon stem-loop tRNA*, and the third key, *single-stranded (GNC)$_n$ RNA*, required only the first key and both the first and the second keys, respectively. The fact indicates also here that the first life could emerge on the primitive Earth, according to not "gene/replicator-first theory" but "protein/metabolism-first theory" (Ikehara 2018).
(4) Fig. 9.10 also shows that the first key or immature [GADV]-proteins were produced using [GADV]-amino acids, which could be synthesized with prebiotic means more easily than nucleotides and RNA. On the contrary, AntiC-SL tRNAs and ss-$(GNC)_n$ RNA were produced with more complex nucleotides, which is difficult to synthesize prebiotically on the primitive Earth, than [GADV]-amino acids.

Table 9.3 All of three materials, a modern mature protein composed of 20 amino acids, a modern L-form tRNA, and a modern gene encoding a mature protein, cannot be created through random processes. However, (1) an immature protein can be produced by random joining of [GADV]-amino acids. (2) AntiC-SL tRNA can be produced through random process assisted by the immature proteins. (3) ds-$(GNC)_n$ gene can be created through ss-$(GNC)_n$ RNA, which was produced through random process assisted by immature proteins and AntiC-SL tRNAs. Thus, a hierarchy can be seen in formation processes of the three keys

Modern member	**m-Protein**	**L-form tRNA**	**Modern gene**
Random process	–	–	–
Primitive member	**imm-Protein**	**AntiC-SLtRNA**	**ds-$(GNC)_n$ gene**
Random process	+ (directly)	+ (indirectly)	+ (indirectly)

(5) Although it is impossible to replicate all the three keys themselves, which were formed by random reactions, only ds-$(GNC)_n$ gene, which was created with the three keys, can be replicated through non-random process.
(6) The first ds-AntiC-SL tRNA gene was created earlier than the first gene for a mature [GADV]-protein (Fig. 9.8a).
(7) Both a mature gene and a mature protein with ordered sequence must be always created from the respectively corresponding immature gene and immature protein. Stating this inversely, it would be impossible to produce both a mature gene and a mature protein directly by random joining of the respective monomeric units, mononucleotides and amino acids.

Thus, two factors made it possible to establish the first fundamental life system in the absence of gene. The first one is protein 0th-order structure (immature [GADV]-protein) and the second one is the selection of [GADV]-microsphere with a higher proliferation ability based on a probability distribution of the microspheres (selection and evolution of [GADV]-microsphere) (Chap. 4: Fig. 4.7). Thus, immature proteins, which were produced by random joining of [GADV]-amino acids, could synthesize nucleotides, oligonucleotides and RNA (Chap. 5: Sect. 3.1.3). Therefore, the greatest difference between my idea and those of other researchers would be attributed simply by whether I fortunately recognized the two factors, or other researchers unfortunately did not.

Strengths of GADV Hypothesis It can be concluded according to the necessary conditions for the origin of something that GADV hypothesis is valid as the origin of life, because it can be reasonably explained like that all the six members of the fundamental life system were formed through the respective random processes and that the respective six primitive ones have evolved to the respective modern ones, as described below (Fig. 9.11).

1. The origins and evolutionary processes of the six members can be comprehensively explained according to the GADV hypothesis without any large discrepancy.
2. Formation processes of the six members can be explained under a common concept, protein 0th-order structure.

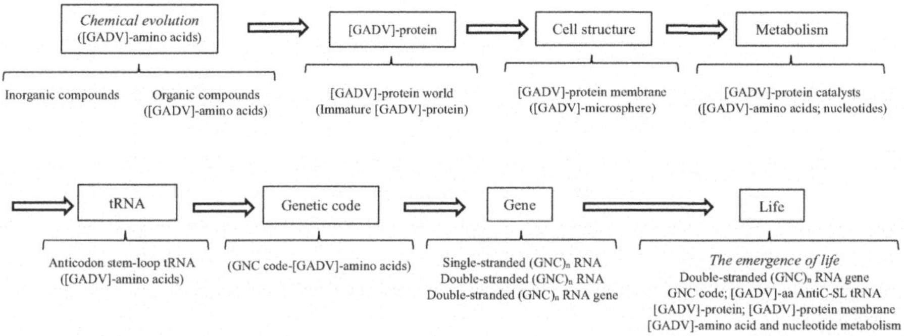

Fig. 9.11 The steps from chemical evolution to the emergence of life, which were obtained mainly with data analyses of amino acid sequences of proteins and base sequences of genes in extant microorganisms. Only main points of the GADV hypothesis are summarized in the figure. Note that at least one of [GADV]-amino acids, [GADV]-proteins, GNC code encoding [GADV]-amino acids and $(GNC)_n$ gene encoding [GADV]-protein are contained in all eight items

3. In addition, GADV hypothesis also satisfies the condition that the evolutionary processes from the six most primitive members to the respective modern ones must be explained (Chaps. 3, 4, 5, 6, 7 and 8).
4. Evolutionary processes of the six members can be also explained by coevolution among the six members.
5. Many matters included in GADV hypothesis are confirmed by analyses of databases of microbial genes, tRNAs and proteins, which were obtained by experiments.
6. Some matters as catalytic activities of aggregates of [GADV]-peptides and formation of [GADV]-microspheres are confirmed by experiments.

Weaknesses of GADV Hypothesis

1. As a matter of course, it cannot be always concluded that the first life emerged according to the steps deduced from GADV hypothesis on the grounds that the evolutionary process from chemical evolution to the emergence of life can be reasonably explained by the hypothesis.

   Similarly, it cannot be also concluded that the evolutionary process proceeded along with the steps deduced from GADV hypothesis, based on that the evolutionary process could not be reasonably explained by ideas proposed by other researchers.
2. It is the greatest weakness that many steps deduced from GADV hypothesis are not confirmed by experiments so far. Therefore, it is a future problem of GADV hypothesis, whether the steps deduced from the hypothesis could be confirmed by experiments or not.

5 Discussion

I have discussed in detail the origins of the six members composing the fundamental life system in the respective chapters from 3 to 8. Here, I discuss further my idea about the origin of life under the basis of the origins of the six members.

Then, the reasons might become a problem anew, why many researchers could not solve the riddle of the origin of life, irrespective of their strenuous efforts. The reasons are considered in this section as adding new knowledge obtained through considerations about the origins of the six members of the fundamental life system as follows.

1. Researchers, who want to solve the riddle of the origin of life, will usually investigate from the viewpoint of bottom-up approach trying to clarify what happened on the primitive Earth and what kinds of organic compounds were produced.
2. It seems to me that, accordingly, many studies on the origin of life were carried out by other researchers thus far without considering the way how the fundamental life system was established on the primitive Earth.
3. However, it would be impossible to make clear the riddle of the origin of life without considering the most important point of the establishment process of the fundamental life system.
4. Therefore, many researchers might tend to consider the origins of the six members including gene, tRNA (genetic code) and protein, composing the core life system, not comprehensively but individually (Table 9.2).
5. Furthermore, only random reactions could occur on the primitive Earth before the first gene was created. It would be difficult to clarify the origins of the three main members of gene, tRNA (genetic code) and protein without considering whether or not something assumed as the origin could be produced under random reactions.
6. Therefore, they could not find out the concept of "protein 0th-order structure", in which not mature but immature [GADV]-proteins could be produced by random joining of [GADV]-amino acids.
7. Furthermore, many researchers might overlook the point of view whether or not something assumed as the origin could evolve to the modern one.
8. It seems to me that many researches have paid much attention to the features of the respective modern members constituting the fundamental life system to make clear the origin of life. Such approaches seems to be apparently reasonable. However, the trials would not succeed if the steps to the emergence of life had proceeded as piling up the members one by one, as expected by GADV hypothesis (Fig. 9.11). I have supposed that such origin-life researches trying to individually clarify the origin of life instead might make it difficult to solve the origin of life.

On the contrary, I fortunately reached the GADV hypothesis. It would be important to consider the reasons why I could reach the GADV hypothesis. The reasons, which I consider, are as follows.

1. I have studied rather theoretically on the origin of life as analyzing databases of microbial genes, proteins and so on. I feel now that analyses of the databases led me to the GADV hypothesis.
2. I have accidentally explored the study on the origin of life through a top-down approach, as I started with my studies from a question how and from where an EntNew gene has been created on the present Earth.
3. Consequently, I could reach the novel establishment process of the fundamental life system in order of the origins of protein, genetic code and gene, owing to the top-down approach.
4. Furthermore, I could reach the novel idea, which can explain the six origins of protein, cell structure, metabolism, tRNA, genetic code and gene comprehensively. Nobody would doubt that extant organisms are living under the fundamental life system composed of the six members. In addition, many researchers could agree with the idea that the first life emerged on the primitive Earth upon acquisition of the six members.
5. During the studies, I could fortunately notice a protein 0th-order structure composed of [GADV]-amino acids, in which water-soluble globular protein could be produced even by random joining of [GADV]-amino acids. Inversely, it must be difficult to explore the origin of life without the concept of protein 0th-order structure.
6. It means that I could reach the GADV hypothesis as a trigger of acquisition of GC-NSF(a) hypothesis on EntNew gene creation in the present microorganisms. Another reason is because, as a result, my research has accidentally but fortunately progressed as going up through the core part of the fundamental life system (gene, genetic code and protein).

Thus, I could fortunately meet with protein 0th-order structure as a clue of the GC-NSF(a) hypothesis during my research process. Inversely stating that, I have to state that all my ideas on the origin of life based on the six members or the fundamental life system are wrong, if the protein 0th-order structure were wrong. Therefore, I consider that it would become an important problem in a future whether or not the protein 0th-order structure is correct or wrong.

In addition, as seen in Fig. 9.11, it is confirmed that the six origins were formed under the common concept of protein 0th-order structure or [GADV]-amino acids. It would not be accident that such affairs happened under the common concept as a whole. Inversely stating this, it seems to me correct that the first life emerged as expected by GADV hypothesis. Of course, many readers may consider that it is natural and almost meaningless for me to endorse the idea, GADV hypothesis, which I myself have proposed. However, it seems to me still that the steps to the emergence of life proceeded according to the GADV hypothesis, even if I take it into consideration that I have assessed my own idea.

The Steps from Immature [GADV]-Proteins Produced with Prebiotic Means to the Emergence of Life

Finally, I summarize the steps from immature [GADV]-proteins produced with prebiotic means to the emergence of life, which I have deduced according to GADV

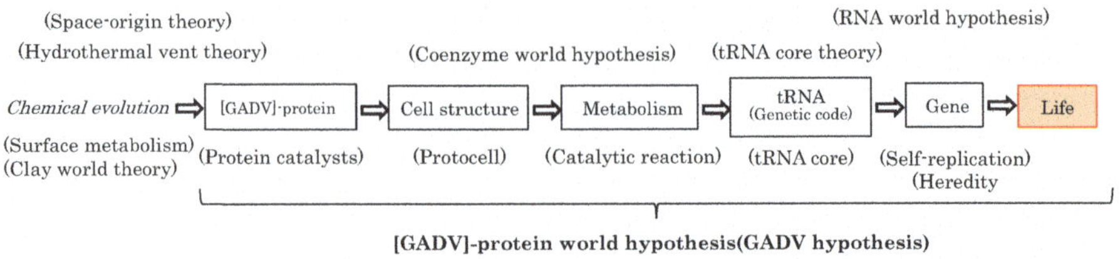

Fig. 9.12 The six steps (bold white arrows) from [GADV]-protein to the emergence of life (red box) deduced from [GADV]-protein world hypothesis or GADV hypothesis (bold letters). As can be seen in the figure, other theories and hypotheses seem to try to clarify some matters described in parenthesis on the way to the emergence of life. The whole origins of the six members cannot be explained with the ideas, which have been proposed by other researchers. The evolutionary steps from the primitive one produced through random process to the modern one have not been concretely presented by other researchers. Contrary to that, evolutionary process, from the six members produced through the respective random processes to modern ones, could be reasonably explained by GADV hypothesis, which I have proposed

hypothesis, as comparing with the ideas on the origin of life proposed by other researchers, as shown in Fig. 9.12.

As can be seen in Fig. 9.12, many other researchers, which have presented the theories or hypotheses on the origin of life, have paid attention rather to individual members constituting the fundamental life system. It would be quite difficult to solve the origin of life by their ideas, because the solution should be obtained through understanding the whole life system comprehensively. On the contrary, the origin of life has been explored in the GADV hypothesis through understanding the whole six members, and, therefore, all indispensable factors have been contained in the steps from chemical evolution to the emergence of life, which are deduced from GADV hypothesis.

As described by Stanley L. Miller and Leslie E. Orgel in their book (1974) as follows, as I showed the same paragraph in Chap. 3.

> Amino acids were the first biologically interesting organic compounds to be identified as products formed under simulated primitive earth conditions. This was partly because amino acids were more easily identified at the time than purines and pyrimidines, and partly because they were formed in a single continuous set of operations under reasonably plausible prebiotic conditions.

The description is quite natural, because life should emerge on the primitive Earth using organic compounds, which could be produced with prebiotic means and accumulated at a large amount on the primitive Earth.

Further, Fox described in the paper published from his group, as I showed the same paragraph in Chap. 4.

> Among the contributors whose ideas can be interpreted as consonant with that of a cell initiating protein and nucleic acid synthesis and Darwinian selection have been Wald (1954), Van Niel (1956), Oparin (1957), Lederberg (1959), Ehrensvard (1960-1962), and Prosser (1970). The integrated view from their hypotheses and from our experimental results is expressed concisely by the sequence: protoprotein – heterotrophic protocell – macromolecule-synthesizing cell. (Fox et al. 1974)

The idea described by Fox et al is also natural as the steps to the emergence of life.

I have explained according to GADV hypothesis in this book, about how the first life emerged on the primitive Earth. Next, consider the reason why life could emerge on this planet.

Many key matters contributed to the emergence of life as described below.

(1) [GADV]-amino acids with a quite simple chemical structure existed in this world. Immature [GADV]-proteins, actually aggregates of [GADV]-peptides, could be produced probably by repeated wet-drying cycles in depressions of rocks on the primitive Earth owing to the existence of [GADV]-amino acids.
(2) [GADV]-microsphere could be formed with the immature [GADV]-proteins.
(3) Various metabolic reactions producing [GADV]-amino acids, [GADV]-peptides, nucleotides and RNA etc., could be catalyzed by immature [GADV]-proteins in the [GADV]-microsphere or in [GADV]-protein world.
(4) The first AntiC-SL tRNA could be formed by repeated RNA synthesis and its degradation cycles with immature [GADV]-proteins.
(5) The primeval GNC genetic code could be established frozen accidentally through the formation of the first four specific AntiC-SL tRNAs.
(6) The first single-stranded $(GNC)_n$ gene could be produced by random joining of anticodons carried by AntiC-SL tRNAs.

All the respective steps proceeded as miracles, as if those were prepared previously by something. Above all, the existence of [GADV]-amino acids on the primitive Earth and the formation of the four nonspecific AntiC-SL tRNAs in [GADV]-microsphere played central roles in the emergence of life. The reason is because [GADV]-protein, which was produced with [GADV]-amino acids lead to the formation of the four nonspecific AntiC-SL tRNAs followed by the establishment of the GNC code and production of the first ds-$(GNC)_n$ gene.

6 Conclusion

One of the objectives of the study on the origin of life is to pursue the simplest and most primitive thing with similar features to the modern life. Therefore, it is expected to understand the essence of life through the origin-life research. My rough picture of life, which was obtained from the study on the origin of life, is as follows.

The fundamental features, which even the first life must have, are as follows.

1. Life must be a substance, which is surrounded by membrane and is discriminate from others.
2. Life must be able to grow, proliferate and evolve.

Therefore, the essence of life would be individuality, growth, proliferation and evolvability. Note that self-replication is not included in the conditions viewed from the the essence of life.

Life can be divided into two types, protocell without ds-RNA/DNA, and genuine life with ds-RNA/DNA gene, considering based on the above conditions (see also Chap. 4: Sect. 3.2).

(1) *The first protolife was [GADV]-microsphere without the genetic system*: The reason, why [GADV]-microsphere can be considered as a protocell, is because even the [GADV]-microsphere, which was composed of only [GADV]-proteins actually aggregates of [GADV]-peptides, could grow and proliferate owing to catalytic functions of [GADV]-proteins. In addition, the microsphere enclosed by [GADV]-protein membrane had an individuality, which enabled to evolve through selection among others (Chap. 4: Fig. 4.7).

It would be obvious from the following example that [GADV]-microsphere could evolve. Even simple chemical compounds were gradually but continuously complicated from inorganic compounds to organic compounds in the era of chemical evolution. The reason is because chemical equilibrium between inorganic compounds and organic compounds always constantly proceeds from inorganic compounds previously accumulated at a large amount to organic compounds, which did not exist on the primitive Earth. Organic compounds could be further complicated by catalytic reactions of immature [GADV]-proteins, actually aggregates of [GADV]-peptides, at a higher rate than before, after the immature [GADV]-proteins could be produced.

Thereafter, the first but prototype of life was generated when the first cell structure or [GADV]-microsphere was formed, because individuality could be acquired accompanied by the formation of the [GADV]-microsphere enclosed by [GADV]-protein membrane. The acquisition of the individuality made it possible to differentiate and to select [GADV]-microspheres growing at a higher rate from others. Thus, the microspheres with a higher proliferation ability were selected from others (Fig. 9.13). Thus, [GADV]-microspheres evolved gradually toward the microspheres with higher functions in spite of the absence of gene. This involves deeply the discussion about "What is life?", because the cell structure without any gene but with abilities of proliferation and evolution could be called as the first protolife.

During the evolution of the first life, nucleotides and RNA could be synthesized in [GADV]-microspheres or in [GADV]-protein world. Upon acquisition of several ds-$(GNC)_n$ RNA genes, the second life or a genuine cell with genetic materials emerged (Fig. 9.13).

(2) *The second life was a genuine cell with a genetic system*: The genuine life is a growing, proliferating and evolving substance under the fundamental life system composed of the six members or protein, cell structure, metabolism, tRNA, genetic code and gene, which were established as supposed by GADV hypothesis. Therefore, the genuine cell had equipped with the fundamental life system at the time point when the cell emerged.

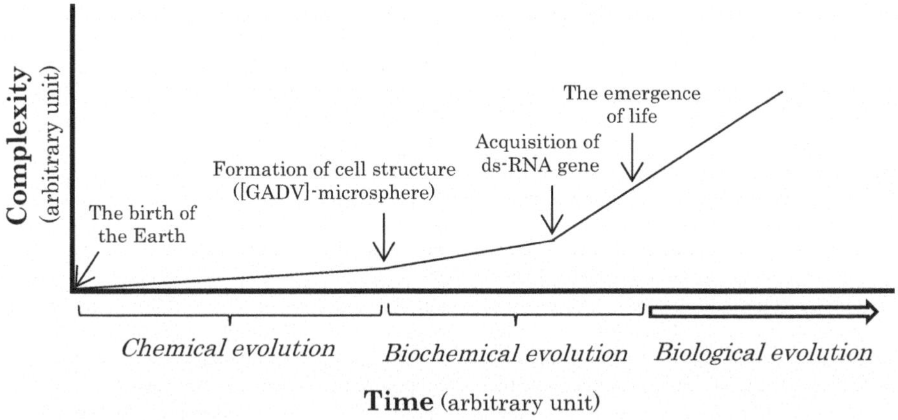

Fig. 9.13 Did life emerged twice? According to GADV hypothesis, it is considered that life emerged twice on the primitive Earth. The first life was a [GADV]-microsphere or preliminary protocell structure, which was composed of [GADV]-proteins or aggregates of [GADV]-peptides, and the second life was a genuine cell equipped with the six members consisting the fundamental life system. Increasing rate of the complexity was accelerated a little accompanied by the formation of the first cell structure. Note that the complexity increased gradually even during chemical evolution. The acquisition of the first ds-RNA gene accelerated further the increasing rate of the complexity much faster than before However, the first genuine life emerged a little later than the acquisition of the first ds-RNA gene or after several ds-RNA genes could be acquired

Of course, the evolution of cell after acquisition of the genetic system including gene is both qualitatively and quantitatively different from the cell structure without gene (Fig. 9.13). Various cell functions could be optimized in the genuine cell, because the cell, which had acquired the genetic system including of ds-$(GNC)_n$-genes, could optimize various cell functions including various protein functions through memorizing ability of ds-$(GNC)_n$-genes.

The genuine life or organism with the genetic system, which evolved from the [GADV]-microsphere, could grow faster than the [GADV]-microsphere without gene (Fig. 9.13), and could adapt to the respective environments on the Earth as acquiring and developing new genes/proteins by utilizing the abilities for increasing the number and the size of genes/proteins. Consequently, diverse organisms have inhabited on the present Earth.

References

Corliss JB, Dymond J, Gordon LI et al (1979) Submarine thermal springs on the galapagos rift. Science 203:1073–1083

De Farias ST, Rêgo TG, José MV (2016) tRNA core hypothesis for the transition from the RNA world to the ribonucleoprotein world. Life (Basel) 6(15)

Dill KA (1990) Dominant forces in protein folding. Biochemistry 29:7133–7155

Ellington AD, Szostak JW (1990) In vitro selection of RNA molecules that bind specific ligands. Nature 346:818–822

Fox SW, Jungck JR, Nakashima T (1974) From proteinoid microsphere to contemporary cell: formation of internucleotide and peptide bonds by proteinoid particles. Orig Life 5:227–237

Furukawa Y, Chikaraishi Y, Ohkouchi N et al (2019) Extraterrestrial ribose and other sugars in primitive meteorites. Proc Natl Acad Sci 116:24440–24445

Gilbert W (1986) Origin of life: the RNA world. Nature 319:618

Gorlero M, Wieczorek R, Adamala K et al (2009) Ser-His catalyses the formation of peptides and PNAs. FEBS Lett 583:153–156

Guerrier-Takada C, Gardiner K, Marsh T et al (1983) The RNA moiety of ribonuclease P is catalytic subunit of the enzyme. Cell 35:849–857

Holm NG, Andersson E (2005) Hydrothermal simulation experiments as a tool for studies of the origin of life on Earth and other terrestrial planets: A review. Astrobiology 5:444–460

Ikehara K (2002) Origins of gene, genetic code, protein and life: comprehensive view of life systems from a GNC-SNS primitive genetic code hypothesis. J Biosci 27:165–186

Ikehara K (2005) Possible steps to the emergence of life: The [GADV]-protein world hypothesis. Chem Rec 5:107–118

Ikehara K (2009) Pseudo-replication of [GADV]-proteins and origin of life. Int J Mol Sci 10:1525–1537

Ikehara K (2014) Protein ordered sequences are formed by random joining of amino acids in protein 0th-order structure, followed by evolutionary process. Orig Life Evol Biosph 44:279–281

Ikehara K (2017) GADV-protein: an alternative to RNA in the origin of life. Preprints. MDPI, Basel. https://doi.org/10.20944/preprints201712.0170.v1

Ikehara K (2018) Life emerged as the "protein/metabolism-first" theory expects. Preprints. MDPI, Basel. https://doi.org/10.20944/preprints201811.0620.v1

Ikehara K (2019) The Origin of tRNA deduced from *Pseudomonas aeruginosa* 5′anticodon-stem sequence -anticodon-stem loop hypothesis. Orig Life Evol Biosph 49:61–75

Ikehara K, Yoshida S (1988) SNS hypothesis on the origin of the genetic code. Viva Orig 26:301–310

Ikehara K, Amada F, Yoshida S et al (1996) A possible origin of newly-born bacterial genes: Significance of GC-rich nonstop frame on antisense strand. Nucleic Acids Res 24:4249–4255

Ikehara K, Omori Y, Arai R et al (2002) A novel theory on the origin of the genetic code: a GNC-SNS hypothesis. J Mol Evol 54:530–538

Jacobsen NE, Pfaltz A, Yamamoto H (eds) (1999) Comprehensive asymmetric catalysis. Springer, New York

Kruger K, Grabowski PJ, Zaug AJ et al (1982) Self-splicing RNA: autoexcision and autocyclization of ribosomal RNA intervening sequence of *Tetrahymena*. Cell 31:147–157

Lee DH, Granja JR, Martinez JA et al (1996) A self-replicating peptide. Nature 382:525–528

Li X, Chmielewski J (2003) Peptide self-replication enhanced by a proline kink. J Am Chem Soc 125:11820–11821

Luisi PL (2014) A new start from ground zero? Orig Life Evol Biosph 44:303–306

Luisi PL (2016) The emergence of life-from chemical origins to synthetic biology, 2nd edn. Cambridge University Press, Cambridge

Maury CPJ (2009) Self-propagating β-sheet polypeptide structures as prebiotic informational molecular entities: the amyloid world. Orig Life Evol Biosph 39:141–150

Maury CPJ (2018) Amyloid and the origin of life: self-replicating catalytic amyloids as prebiotic informational and protometabolic entities. Cell Mol Life Sci 75:1499–1507

Miller SL, Orgel LE (1974) The origins of life on the earth. Prentice-Hall, Englewood Cliffs

Oba T, Fukushima J, Maruyama M et al (2005) Catalytic activities of [GADV]-peptides. Formation and establishment of [GADV]-protein world for the emergence of life. Orig Life Evol Biosph 35:447–460

Sharov AA (2009) Coenzyme autocatalytic network on the surface of oil microspheres as a model for the origin of life. Int J Mol Sci 10:1838–1852

Sharov AA (2016) Coenzyme world model of the origin of life. Biosystems 144:8–17. https://doi.org/10.1016/j.biosystems.2016.03.003
Shibasaki M, Yoshikawa N, Matsunaga S (2004) In: Jacobsen NE, Pfaltz A, Yamamoto H (eds) Comprehensive asymmetric catalysis, vol 1. Springer, New York, pp 135–142
Suwannachot Y, Rode BM (1998) Catalysis of dialanine formation by glycine in the salt-induced peptide formation. Orig Life Evol Biosph 28:79–90
Thodima V, Pirooznia M, Deng Y (2006) RiboaptDB: a comprehensive database of ribozymes and aptamers. BMC Bioinform 7(Suppl 2):S6
Tuerk C, Gold L (1990) Systematic evolution of ligands by exponential enrichment: RNA ligands to bacteriophage T4 DNA polymerase. Science 249:505–510
Van der Gulik P, Massar S, Gilis D et al (2009) The first peptides: the evolutionary transition between prebiotic amino acids and early proteins. J Theor Biol 261:531–539
Watson JD, Crick FHC (1953) Molecular structure of nucleic acids: a structure for deoxyribose nucleic acid. Nature 171:737–738
Zhang X, Tian G, Gao J et al (2017) Prebiotic synthesis of glycine from ethanolamine in simulated Archean alkaline hydrothermal vents. Orig Life Evol Biosph 47:413–425

Chapter 10
General Discussion

Abstract The matters, which are described rather individually, are discussed comprehensively from three viewpoints in the last Chap. 10. First, establishment process of the six members constituting the fundamental life system is summarized based on the discussions described in the respective chapters. Successively, two research approaches exploring the origins of six members constituting the fundamental life system or protein, cell structure, metabolism, tRNA, the genetic code and gene. One approach is first to explore from the origin of one member, and the origins of remaining five members are made clear one by one in order as a start point of the origin of the first member. In this case, it is expected that the origins of the whole members could be made clear, if the studies had started from the origin of the right member and if the origins of all the six members had related to each other as tying in a row. Fortunately, GADV hypothesis was obtained according to the first strategy. The second one is to start from studies on the origins of a plural number of members followed by unifying the origins of the whole members, which were made clear individually. However, in the latter case, in the latter case, it would be difficult to unify the six origins without any discrepancy, because the respective different components should be generally assumed to study the origins of the six members and discrepancies among the origins would appear at a high probability. Finally, evolutionary processes from the respective origins followed by prosperity of modern organisms on the present Earth are discussed.

Keywords Top-down approach · Evolution of organisms

1 What is the Significance of the Origin-Life Research?

The studies on the origin of something including the origin of life is to explore the events, which happened on the primitive Earth about 4 billions years ago. Therefore, the origin-life research must treat the events, which would not be reproduced and, therefore, it is difficult to confirm with experiments in a present laboratory. One day, I have heard someone saying that the origin of life should not be a theme, which decent "scientists" tackle, and ask only "How", don't ask "Why". Is that correct? If

K. Ikehara, *Towards Revealing the Origin of Life*,
https://doi.org/10.1007/978-3-030-71087-3_10

so, we cannot ask "Why did life emerge on this planet?" The attitude is to give up knowing the creation mechanism of the fundamental life system. It would be true that it often seems to be impossible to solve it, when researchers face a quite difficult problem and until someone solves the problem. On the other hand, it is the unquestionable fact that organisms, which emerged about 4 billion years ago on this planet, inhabit on the present Earth. Therefore, the origin of life is a problem, of which the riddle must be solved someday, although the origin of life is certainly one of the quite difficult problems.

I have described in Chap. 1 of this book that there are serious questions and problems in the origin-life research as "Why has not the origin of life been solved thus far?", "The reasons from viewpoints of research methodologies" and so on. Further, I have also described "Towards solving the riddle of the origin of life", "Necessary matters for solving the riddle of the origin of life" and "Significance of theoretical approaches" in Chap. 2. I have indicated the significance of recognizing the features of the six modern members composing the fundamental life system and of confirming whether something assumed as the origin of a member could evolve to the modern one or not. I have also indicated in this book that the reason, why the origin of life could not be solved, might be because such problems above have not been sufficiently considered.

I have studied on the origin of life as considering the important points for revealing the origin of life. From the results, I have felt like that I could understand, how all the so beautiful six members constituting the fundamental life system were created, and that I might solve the riddle of the origin of life, although I did not start aiming at solving the origin-life research at first (Chap. 1: Fig. 1.2).

2 Establishment Process of the Fundamental Life System

All members constituting the fundamental life system have the respective splendid features, because all extant organisms have evolved from the first small but wonderful life with an evolvability over about 4 billion years on this Earth. On the other hand, all the members must be generated either directly or indirectly through random processes. It could be understood according to the GADV hypothesis, the way how the amazing fundamental life system could be established during such random processes. [GADV]-proteins produced under a protein 0th-order structure triggered to establish the fundamental life system (Fig. 10.1).

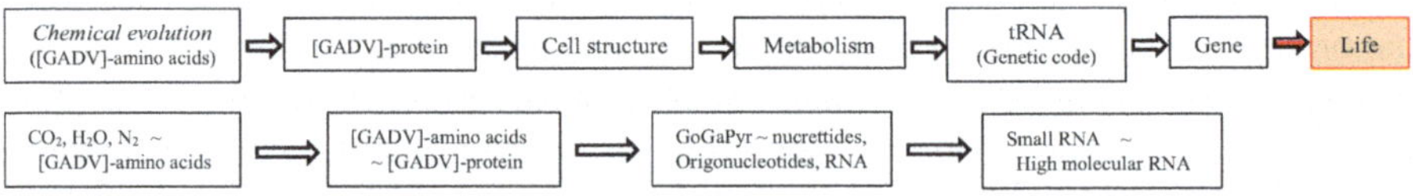

Fig. 10.1 The evolutionary steps from [GADV]-amino acid synthesis during chemical evolution to the emergence of life, which are assumed by GADV hypothesis. As can be seen in the lower evolutionary steps, the formation processes of the six members proceeded from productions of more simple organic compounds ([GADV]-amino acids) towards syntheses more complex organic compounds (nucleotides and RNA)

Then, discuss the formation process of the systematized fundamental life system in more detail. As easily supposed, two ways could be considered about the formation process. One way is to form the whole system by starting from one member, which is simple and primitive as much as possible. Therefore, it would be easy to synthesize the member on the primitive Earth. The formation process of the fundamental life system spread gradually to other members, which are more complex and are more difficult to produce, using functions of the firstly produced member. It would be the most reasonable to complete the fundamental life system as piling up the remaining five members one by one on the basis of the first member.

The formation process deduced from GADV hypothesis indicates that the fundamental life system was formed as piling up the members in order of cell structure, metabolism, tRNA, genetic code and gene on the basis of protein or immature [GADV]-protein one by one, as shown in Fig. 10.1 (see also Chap. 9: Fig. 9.5).

3 About Another Possibility of the Establishment Process of the Fundamental Life System

As described above, (1) one way for formation of the whole fundamental system is to start from one member and to spread gradually to other members. (2) Another way might be also possible. That is to begin the formation process from a plural number of members independently and to unify all the members finally.

Then, consider that first some members were formed through random processes independently. However, such an establishment process of the life system should meet some quite difficult problems as described below (Fig. 10.2).

(1) Formation of a plural number of members, itself, would be disadvantage.
(2) The reason is because it should be difficult to produce even one member through random process. Therefore, it would be more difficult to produce a plural number of the members through random processes.

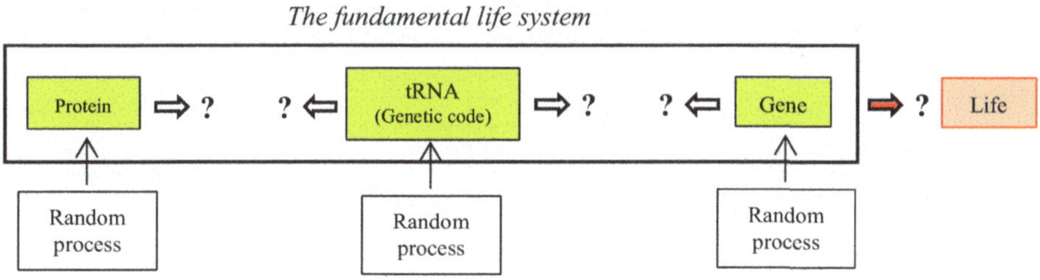

Fig. 10.2 For example, assume that three members or protein, tRNA and gene, were produced independently through the respective random processes. Then, three independent random processes are required to form the three primitive members. It would be difficult to form the three members by three random processes. In addition, the three primitive members obtained must be unified to establish the fundamental life system. However, it would be also quite difficult to unify the three members, which were formed independently, because the respective members should be produced with different components

(3) It is further difficult to establish the fundamental life system by connecting a plural number of members, which were formed independently. The reason is because the respective members are generally different from each other and, therefore, it should be difficult to connect with the members as seen in Fig. 10.2.

As easily understood, it would be easy to establish the system according to the former strategy described in Sect. 2. The GADV hypothesis fortunately follows the strategy (Fig. 10.1).

4 Bottom-Up and Top-Down Approaches for Elucidation of the Origin of Life

The research on the origin of life is generally carried out with the bottom-up approach. However, it would be difficult to clarify all steps, which should be composed of many steps from chemical evolution to the emergence of life, only with the bottom-up approach (Fig. 10.3). Fortunately, GADV hypothesis, which I

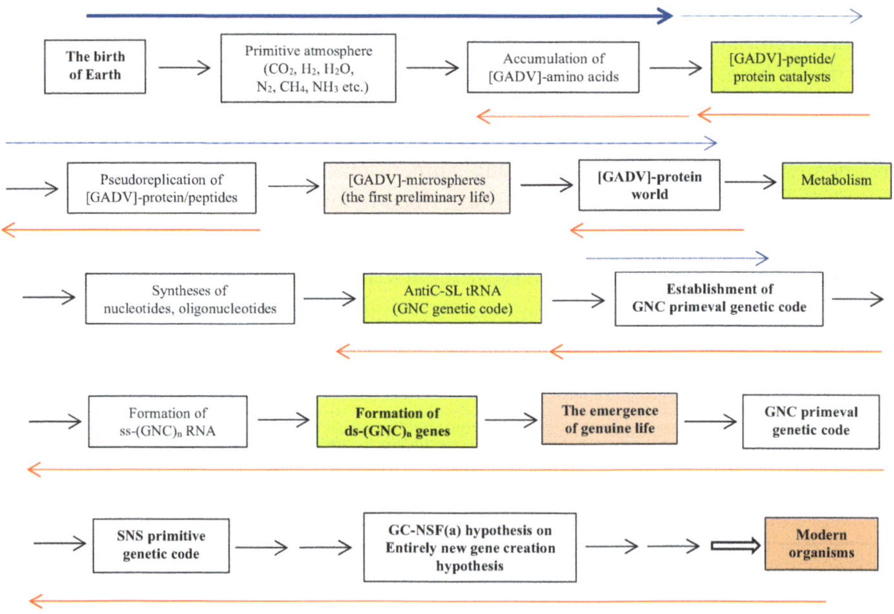

Fig. 10.3 The evolutionary steps from the birth of Earth to the modern organisms, which are deduced by connecting the two results obtained by bottom-up approach (upper blue right arrows) and top-down approach (lower red left arrows). The main evolutionary steps, which were deduced by GADV hypothesis, are drawn in the figure. Bold arrows and thin arrows show the steps deduced from the two approaches with a higher and a lower reliability, respectively. It can bee seen in the figure that the number of steps, which were deduced from the bottom-up approach, is smaller than the number of steps deduced from the top-down approach. It can be also understood from the figure that there are not large discrepancies in the whole evolutionary steps, which are deduced from the two results obtained by bottom-up and top-down approaches

have proposed, is the idea, which was obtained with top-down approach. Therefore, the consecutive evolutionary steps from the birth of Earth to the emergence of life could be rationally deduced using a common event (for example, the establishment of the GNC primeval genetic code) as a juncture for connecting the results obtained from the two counter-directional (bottom-up and top-down) approaches (Fig. 10.3) (Ikehara 2016).

5 Did the Events Deduced from GADV Hypothesis Really Occur on the Primitive Earth?

Many persons may doubt that such events, which are deduced from GADV hypothesis, did really occur on the primitive Earth. Many persons may expect that all the events including the emergence of life and the establishment of the fundamental life system could be explained, if genes could be produced with prebiotic means at first. The reasons, why many persons consider so, are probably because they are fixed on the gene-centered idea, as almost all events in extant organisms have occurred under control of the genetic expression.

Of course, I also consider that all the events could be explained, if genes were produced with prebiotic means at first. However, it would be impossible in principle to produce genes or RNA with prebiotic means at the very beginning, because the main three members of the fundamental life system must be established in order of protein (metabolism), tRNA (genetic code) and gene (Figs. 10.2 and 10.3) and because the life system exists to synthesize protein, and, therefore, both gene and tRNA cannot be created in the absence of protein, and gene cannot be created without protein and tRNA, as described in detail in this book (Fig. 10.2).

6 Evolution of Organisms

The first genuine genetic system ($(GNC)_n$ gene; [GADV]-AntiC-SL tRNAs (GNC code); [GADV]-protein) was established upon the formation of the first ds-$(GNC)_n$ gene. After the establishment of the first genetic system, the first genuine life emerged. Formation of a new genetic sequence by base substitution might trigger formation of new codon on sense strand or on antisense strand of GC-rich gene. However, it would be impossible to use the newly appeared codon for protein synthesis in the absence of tRNA translating the codon. Therefore, it is considered that a new amino acid, which was produced through development of a new amino acid synthetic pathway, should trigger formation of a new tRNA (genetic code) and gene (genetic information). Therefore, accumulation of a new amino acid in cells induced formations of a new tRNA (genetic code), new genes and new proteins in parallel (Fig. 10.4). Biological evolution after the emergence of the first genuine life proceeded as enhancing functionality of essentially the same life system with the most primitive life system.

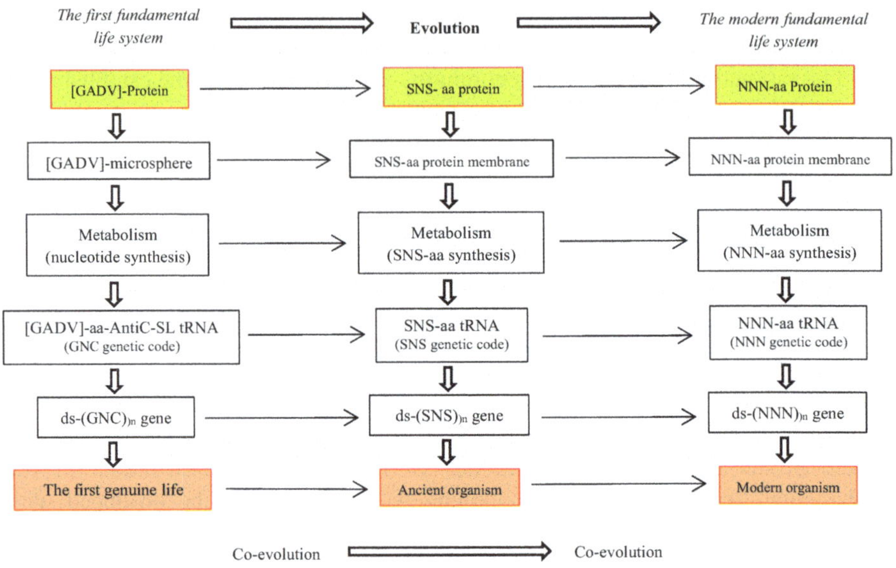

Fig. 10.4 After establishment of the fundamental life system composed of the six members or protein, cell structure, metabolism, tRNA (genetic code) and gene, the most primitive life evolved to modern organisms through ancient organisms lived under the genetic system composed of $(SNS)_n$ gene, SNS code and proteins using 10 amino acids encoded by the SNS code. Evolution of the life system would be always initiated by development of a new amino acid synthetic pathway

7 Why Have So Splendid Organisms Flourished on the Present Earth?

The reasons are because extant organisms have evolved based on the following fulfilling basis.

1. The first one is the protein 0th-order structure or [GADV]-amino acids and [GADV]-protein composed of [GADV]-amino acids. Amazing modern proteins, which are composed of 20 amino acids frequently called as "magic 20", could be formed on the basis of [GADV]-amino acids, which have themselves at high levels, as adding talented amino acids one by one.
2. The second one is a sufficiently stable [GADV]-microsphere having robust viability, which was composed of [GADV]-proteins. The [GADV]-microsphere could incorporate and concentrate various organic compounds accumulated in surroundings. Progenies of the first [GADV]-microsphere could proliferate at a higher rate than before, as feeding nonproductive [GADV]-microspheres too, which stored up many organic compounds like as [GADV]-proteins.

The six primitive members have evolve and have been highly developed over about 4 billion years to the respective modern members on the two bases of [GADV]-amino acids and [GADV]-microsphere (Fig. 10.4). These can be confirmed as seen in figures (Fig. n.1) appeared at the first figure of the respective chapters of this book.

Reference

Ikehara K (2016) Evolutionary steps in the emergence of life deduced from the bottom-up approach and GADV hypothesis (top-down approach). Life 6:6

Index

K. Ikehara, *Towards Revealing the Origin of Life*,
https://doi.org/10.1007/978-3-030-71087-3

The manufacturer's authorised representative in the EU is Springer
Nature Customer Service Centre GmbH, Europaplatz 3, 69115 Heidelberg,
Germany. If you have any concerns regarding our products, please
contact ProductSafety@springernature.com

Printed and bound by CPI Group (UK) Ltd, Croydon, CR0 4YY
30/09/2026
02225691-0002